About Island Press

Island Press is the only nonprofit organization in the United States whose principal purpose is the publication of books on environmental issues and natural resource management. We provide solutions-oriented information to professionals, public officials, business and community leaders, and concerned citizens who are shaping responses to environmental problems.

Since 1984, Island Press has been the leading provider of timely and practical books that take a multidisciplinary approach to critical environmental concerns. Our growing list of titles reflects our commitment to bringing the best of an expanding body of literature to the environmental community throughout North America and the world.

Support for Island Press is provided by the Agua Fund, The Geraldine R. Dodge Foundation, Doris Duke Charitable Foundation, The Ford Foundation, The William and Flora Hewlett Foundation, The Joyce Foundation, Kendeda Sustainability Fund of the Tides Foundation, The Forrest & Frances Lattner Foundation, The Henry Luce Foundation, The John D. and Catherine T. MacArthur Foundation, The Marisla Foundation, The Andrew W. Mellon Foundation, Gordon and Betty Moore Foundation, The Curtis and Edith Munson Foundation, National Fish and Wildlife Foundation, Oak Foundation, The Overbrook Foundation, The David and Lucile Packard Foundation, Wallace Global Fund, The Winslow Foundation, and other generous donors.

The opinions expressed in this book are those of the author(s) and do not necessarily reflect the views of these foundations.

Ecology and
Ecosystem Conservation

Foundations of Contemporary Environmental Studies

James Gustave Speth, editor

Global Environmental Governance
by James Gustave Speth and Peter M. Haas

Ecology and Ecosystem Conservation
by Oswald J. Schmitz

Forthcoming:

Environmental Economics
by Nathaniel Keohane and Sheila Olmstead

Nature and Human Nature
by Stephen Kellert

Environmental Policy and Law
by Daniel Esty and Douglas Kysar

Environmental Health
by John Wargo

ECOLOGY AND ECOSYSTEM CONSERVATION

Oswald J. Schmitz

Yale University School of Forestry
and Environmental Studies

WASHINGTON · COVELO · LONDON

For Coulter, Zachary, and Cameron.
This is my effort to give them a
bright environmental future.

Library of Congress Cataloging-in-Publication Data

Schmitz, Oswald J.
Ecology and ecosystem conservation / by Oswald J. Schmitz.
   p.   cm.
Includes bibliographical references.
ISBN-13: 978-1-59726-048-0 (cloth : alk. paper)
ISBN-10: 1-59726-048-7 (cloth : alk. paper)
ISBN-13: 978-1-59726-049-7 (pbk. : alk. paper)
ISBN-10: 1-59726-049-5 (pbk. : alk. paper)
   1. Ecology.   2. Biodiversity conservation.   I. Title.
QH541.S325 2007
577—dc22   2006035637

Printed on recycled, acid-free paper ⊛

Manufactured in the United States of America

10 9 8 7 6 5 4 3 2 1

Table of Contents

Preface ix

Chapter 1: Ecosystem Conservation: The Need for Ecological Science 1

Chapter 2: The Science of Ecology 6
What is Is Ecology? 7
Resolving Ecological Complexity 7
Life as a Game 15
Ecological Science: Gaining Reliable Knowledge about Ecosystems. 20

Chapter 3: Climate—Template for Global Biodiversity 26
The Physics Underlying Life on Earth 26
Ecosystem Types 28
Coping With with Climate 29
Climate-Space 35
Effects of Global Climate Change 36

Chapter 4: Ecological Limits and the Size of Populations 45
Simple Population Growth 46
Ecological Balance and Carrying Capacity 49
Competitors and Predators 54
Weather 55
Carrying Capacity and Population Overabundance 57

Chapter 5: Viability of Threatened Species 63
Life-Cycles and Population Dynamics 66
Modeling Age-Structured Population Dynamics 68
Viability of Loggerhead Sea Turtles. 75

Chapter 6: Biodiversity and Habitat Fragmentation 79
Diversity Indices 80
Habitat Fragmentation and the Species-Area Relationship 83
Habitat Fragmentation and Population and Community Processes 88

Chapter 7: The Web of Life: Connections in Space and Time   92
Ecosystems in Time   93
Ecosystems in Space: Linkages Across Geographic Boundaries   96

Chapter 8: Ecosystem Services of Biodiversity   102
Diversity Begets Ecosystem Stability   106
Diversity-Productivity Relations   110
Crop Pollination   111
Pest Control   113
Invasion Resistance   114

Chapter 9: Protecting Biological Diversity and Ecosystem Function   115
Conservation Tools   116
Dynamic Landscapes   121
Global Climate Change and Reshuffling of Faunas   124

Chapter 10: The Good of a Species: Toward a Science-Based Ecosystem Conservation Ethic   126
Tinkering with Economies   127
Ecological Science, Uncertainty, and Precaution   130
Policy and Management as a Scientific Enterprise   135

Questions for Discussion   *139*
References   *143*
Further Reading   *149*
Glossary   *153*
About the Author   *157*
Index   *159*

Preface

In writing this book I have taken much inspiration from the writings of ecologist and conservation ethicist Aldo Leopold, who opened his own book—*A Sand County Almanac*—with the declaration "There are some who can live without wild things, and some who cannot. These [writings] are the delights and dilemmas of one who cannot." These words have resonated with me ever since I was first introduced to Leopold's ideas as an undergraduate. Like Leopold, I have a lifelong professional and personal passion to live in, understand, and write about the workings and wonders of the natural world. As an ecologist, I do this for the pure joy of testing out ideas through scientific discovery and reporting on my findings. As an ecologist, I also do this to contribute to a scholarly community whose goal is to build a body of knowledge that can help society make informed decisions about its interactions with the environment.

The prescience that Leopold demonstrated in his writings is remarkable. His "Land Ethic" sketches out many of the modern themes that ecologists address in their research and in their efforts to inform society about ways to reconcile economic development with ecosystem conservation. These include the evolution of species interactions in food webs and the consequences of disrupting this evolutionary process, the interconnectedness of ecosystems in time and space through material flows and species movement, and the services that ecosystems provide to humankind. In the intervening fifty-odd years since *A Sand County Almanac's* publication, the ecological science community has done much to fill in details large and small and thereby provide vibrant color to Leopold's sketches.

I wrote this book to convey these exciting scientific insights to a readership—including undergraduate environmental studies majors and environmental conservation professionals—that is not intimately familiar with ecology as a scientific discipline. My hope is that readers will come to appreciate the intricate ways that humans are connected to their environment and how their interactions can alter the sustainability of the very ecosystems of which they are a part and from which they derive vital services.

I do not consider myself to be an environmentalist, which I define as someone who advocates particular ways of solving problems. As a scientist who studies the workings of ecological systems, I feel it is my duty to present the science as clearly and as objectively as possible, and in ways that illuminate the consequences of different actions so that each reader can make informed decisions about how he or she chooses to interact with the environment. Most importantly, I hope to provide readers the very humbling understanding that the consequences of our decisions today will be felt by our grandchildren and great-grandchildren. These are the timescales—at the least—on which ecosystem functions operate and on which we need to anticipate our impacts.

Several colleagues have provided thoughtful comments and discussion on various drafts, including: Brandon Barton, Holly Jones, Gus Speth, Karen Stamieszkin, and Mark Urban. I very much appreciate the time and effort they put into reviewing. I also thank Leslea Schmitz for her patience during the entire project and for "holding the fort" while I was absorbed in writing.

— Oswald J. Schmitz
Yale University
New Haven, Connecticut

1

Ecosystem Conservation: The Need for Ecological Science

IT IS BECOMING INCREASINGLY IMPOSSIBLE TO TALK ABOUT HUMANS' RELA-tionship to nature without mentioning ecology. More and more, this particular field of science is being called upon to play a leading role in illuminating and solving environmental problems. So much so that the environmental historian Donald Worster suggests that the twenty-first century might well be called the "Age of Ecology" (Worster 1994).

In the post–World War II era, ecological science has played a prominent role in identifying the cause of major environmental problems and motivating consequent policy to mitigate them. Rachel Carson's *Silent Spring* (1961) alerted us to the danger to humans and wildlife species of pesticides, which led to government regulation of chemicals in the environment. The investment of resources and brainpower to discover that phosphate pollution from households caused massive algae blooms that choke out other forms of life in major freshwater bodies (Schindler 1974) was nothing short of an ecological Manhattan project that led to the Clean Water Act. At the same time, the prospect that acid precipitation (Likens and Borman 1974), produced when sulphur and nitrous oxides from industrial and automobile emissions react and mix with atmospheric oxygen and hydrogen and rain back down, could corrode major terrestrial and aquatic ecosystems spurred tougher sulfate and nitrous emissions standards.

Ecological science successfully led to policy solutions to these problems because ecologists could easily trace the causal chain of effects: "the problems could be seen and smelled and their sources easily identified" (Speth 2004). The problems also were localized and they resonated with society because they directly jeopardized local livelihoods and well-being.

Solving other contemporary environmental problems, such as habitat fragmentation and attendant species extinctions (Simberloff and Abele 1976), has been a less successful enterprise. In this case, the solution to the problem—halting land development and massive scale resource extraction—is usually perceived as standing in the way of human enterprise and economic well being. Moreover, those most directly affected by such activity often are non-human species. And, in many cases, the direct consequences of the actions (e.g., tropical forest loss) occur in distant lands under different government regimes. In this case, the problems were "out of [immediate] sight" and so could be relegated "out of mind."

The irony in such reasoning is that we take great pains to understand how one kind of economy—the market economy—functions; and we take great pains to protect the integrity and functioning of the capital markets that drive economic progress. Society spends comparatively much less time thinking about, understanding, and protecting another major economy—the natural economy—resulting from ecosystem functions and services. Like market economies, myriad lines of dependency exist between species of producers and consumers within natural economies. Humans are not exempted from these dependencies. Any collapse in ecosystem functions, including collapse due to deforestation and fragmentation, stands to reverberate through the market economy, in turn, affecting human well being. Therefore, slogans such as "jobs versus the environment" that pit putative economic progress against measures to conserve ecosystem functions may be misguided. Ecosystems ultimately undergird and drive our economic stability.

> Any collapse in ecosystem functions, including collapse due to deforestation and fragmentation, stands to reverberate through the market economy, in turn, affecting human well being. Therefore, slogans such as "jobs versus the environment" that pit putative economic progress against measures to conserve ecosystem functions may be misguided. Ecosystems ultimately undergird and drive our economic stability.

The aim of this book is to offer insight into the link between the diversity of life—biodiversity—and the structure and functioning of ecosystems. As with the problems of the mid 1900s, the role of ecological science is central to identifying and illuminating the intricate ways that nature works. However, unlike in the past, the challenge for ecological science in discerning the causal chain of effects is becoming more difficult. But the challenge is surmountable.

Meeting the challenge requires a new way of thinking about the intricate dependencies between humans and nature in society's endeavor to sustain long-term health and well being.

Human impacts are many, they are global in reach, and they often combine in synergistic or antagonistic ways at many different geographic scales. Thus, the effect of any single impact is often insidious and therefore requires decades to centuries before it becomes fully manifest. It becomes difficult to pinpoint a specific culprit for such ails as rising cancer levels, degradation of water quality, species' limb deformities, endocrine dysfunction, and many others. Answers require in-depth and critical understanding of the complex ways that species and impacts are linked.

Resolving this complexity is what makes ecological science exciting. At the same time, this complexity is what makes environmental problems ecologically "wicked problems" to solve (Ludwig et al. 2001). Murkiness about causality makes it very easy for governments to dismiss a putative cause of any one impact and therefore avoid action to solve the problem. But, is dismissing an environmental problem for lack of clear causal understanding a wise decision? Such a question cannot be answered without first having a clear understanding of the way that impacts propagate along the myriad lines of dependency within ecosystems.

This book aims to offer such understanding by conveying ecological principles that are relevant to the grand scientific questions about sustaining ecosystem functions. In identifying those questions I take some guidance from a forward-looking report produced in the early 1990s on behalf on the Ecological Society of America titled "The Sustainable Biosphere Initiative" (Lubchenco et al. 1991). This report first underscored the point that most of the environmental problems that human society faces are fundamentally ecological in nature.

In anticipation of the increasing need for ecologists to play a leading intellectual role in solving environmental problems, the authors—leading senior ecologists—developed a plan of action to assemble critical scientific knowledge required to con-

> The effect of any single environmental impact is often insidious and therefore requires decades to centuries before it becomes fully manifest. It becomes difficult to pinpoint a specific culprit for such ails as rising cancer levels, degradation of water quality, species' limb deformities, endocrine dysfunction, and many others. Answers require in-depth and critical understanding of the complex ways that species and impacts are linked.

serve and to wisely manage global ecosystems in the twenty-first century. This report recognized that citizens, policy makers, resource managers, and leaders of business and industry routinely must make decisions concerning the exploitation of resources, but that these decisions cannot be made effectively with limited understanding of the interplay between human domination of ecosystems and impacts on ecosystem function.

According to the report, effective environmental decision-making requires better scientific understanding on three major issues at the nexus between human society and their exploitation of ecosystems:

- *Global Change*, which includes the ecological consequences of natural and human-caused changes in climate, soil properties, water quality, and land- and water-use patterns.
- *Biological Diversity*, which includes the natural basis for the distribution and abundance of species and habitats, human-caused alterations to those patterns locally as well as globally, and the link between diversity and the sustainable functioning of ecosystems.
- *Sustainable Ecological Systems*, which includes the response of ecological systems to exploitation and disturbances, the restoration of ecosystems, the sustainable management of ecological systems, and the interface between ecological processes and human social systems.

I deal with each of those issues consistently throughout the book. But each issue can grade into the other. For example, global change through conversion of forest land into agriculture can impact the distribution and abundance of species—biodiversity. Thus, rather than treat each issue separately, they are interwoven throughout book. The Sustainable Biosphere Initiative report also points out that in order to make effective choices and decisions about the environment in light of these issues we need to answer several big questions about ecology and ecological systems. These questions are:

1. What is the role of ecological science in decision-making?
2. What factors govern the assembly of ecosystems and determine their response to various stressors?
3. How does the earth's climate system function and determine the distribution of life on Earth?
4. What factors control the size of populations?

5. What are the population level consequences of species' life-history adaptations?
6. How does fragmentation of the landscape affect the persistence of species on the landscape?
7. How does biological diversity influence ecosystem process?
8. What ecological principles need to be considered in the design of strategies to protect biological diversity?

My aim here is to address these big research questions by structuring the narrative around example environmental problems. At the same time I will show how the questions posed in the Sustainable Biosphere Initiative document have lead to fresh ways of thinking about ecosystems that are directly relevant to solving problems, including the link between biodiversity and ecosystem functioning, valuing ecosystem services, interconnections of ecosystems across geographic scales, and emergence of ecosystem properties consequent to species sorting processes on landscapes.

I deal with each of the questions in individual chapters. The chapters highlight the latest concepts aimed at answering the big research questions. The book then closes with a final chapter that addresses the need, not only to understand ecological science, but to put that science into an ecosystem ethics perspective. It also returns to and answers the question: Is it wise for policy makers to dismiss environmental problems when their cause is uncertain?

In answering this question, I recognize that society must reconcile significant trade-offs between human health and economic welfare and the protection of natural ecosystem function. One role of ecological science, as I see it, is not to judge, but rather to illuminate the ecological consequences of different potential choices that might be made. Another role, which I also try to convey, is to engender new thinking and awareness of the looming spatial and temporal scales of our impact on nature as globalization of market economies increases the human footprint on the environment.

2

The Science of Ecology

ASK SOMEONE TO DESCRIBE AN ECOLOGICAL SYSTEM AND YOU MIGHT GET the response that it is a group of organisms living together in a fixed place. This is a view likely derived from the familiar elementary school science experiment in which soil, water, nutrients such as nitrogen, bacteria, worms, some plants, and perhaps some herbivores such as snails or insects are put into a hermetically sealed glass container, placed in sunlight, and then left to their own devices. Observers of this experiment always marvel that this simple ecosystem is able to maintain itself indefinitely without any kind of nutrient or species input from the outside. This is because the experiment does not merely assemble a haphazard collection of species. Rather, the experiment deliberately assembles species that together create a natural economy involving a chain of production and consumption, albeit of food energy and nutrients, but an economy nonetheless. In this economy, plants draw up water and nutrients from the soil and carbon dioxide from the air and are stimulated by sunlight to convert those different chemicals into tissue; herbivores eat that plant tissue and when old individuals die the chemical constituents of their body are broken down by worms and bacteria and are recycled back through the system. This economy functions whenever the important lines of dependency, that is the linkage between consumers and their resources and the recycling feedbacks, are sustained.

This simple container system is a powerful metaphor for the way species assemble and interact in nature. The processes of production and consumption are fundamental to sustaining the functioning of all ecological systems globally. Natural ecological systems differ from the container system in that they are comprised of vastly more species with many more interdependencies than those found in the glass container. Understanding these complex interdependencies is the fundamental purpose of that subfield of biology known as ecology.

What Is Ecology?

Ecology is a science aimed at understanding:

- The processes by which living organisms interact with each other and with the physical and chemical components of their surrounding environment.
- The way those processes lead to patterns in the geographical distribution and abundance of different kinds of organisms.

The result of the process leading to a pattern is the assembly of a natural economy. In ecology such a natural economy is formally called an ecosystem.

Ecosystems encapsulate many forms of biological diversity (also called biodiversity). Biodiversity results from a variety among individuals comprising a species owing to sex, age, and genetic differences among those individuals. It also stems from differences between species living together in a geographic location. For example, species may differ in their functional roles (e.g., plant, herbivore, carnivore) and the efficiency with which each carries out its function in different environmental conditions. Biodiversity also arises from the myriad ways that species are linked to each other in ecosystems. As a consequence of these many forms of biodiversity, there is considerable complexity underlying the structure of ecosystems. The challenge in ecology is resolving this complexity.

Biodiversity results from a variety among individuals comprising a species due to sex, age, and genetic differences; from differences between species living together in a geographic location; and from the myriad ways that species are linked to each other in ecosystems. As a consequence of these many forms of biodiversity, there is considerable complexity underlying the structure of ecosystems. The challenge in ecology is resolving this complexity.

Resolving Ecological Complexity

One way to begin resolving complexity is to envision an ecosystem as comprised of vertical food chains in which soil nutrients are linked to plants,

plants are linked to herbivores, and herbivores are linked to carnivores. Such linkages indicate that plants are consumers (predators) of soil nutrients, herbivores are consumers (predators) of plants, and carnivores are consumers (predators) of herbivores. Ecologists give such consumer-resource interactions a special name—trophic interactions. Species engaging in a particular kind of trophic interaction belong to the same trophic level of the food chain. So, for example, species engaging in herbivory belong to the herbivore trophic level, species preying on herbivores belong to the carnivore trophic level, and so on.

In addition, plant species are limited by, and thus must compete for, light and soil nutrients. Herbivore species may therefore compete for limited plant resources and carnivores may potentially compete for an even more limited number of herbivores that comprise their prey. Limiting resources and the need to compete for them can lead to ecological innovation in the way species vie for their share of resources. Thus, we can elaborate our vertical conception of an ecological system by envisioning horizontal linkages within a trophic level as species engage in various strategies to maximize consumption of particular resources.

Conceptualizing Predation and Competition

Together, the vertical chain comprised of consumer-resource links coupled with horizontal links between species at the same trophic level create a highly interconnected web of life—a food web. Individual species within this web are sandwiched between their predators, their resources, and their competitors. The easiest way to imagine the implications of such complexity is to begin by drawing food web diagrams that depict the interdependencies among species created by their linkages and the nature of each species' net effect on the other species (figure 2.1). Such an approach assumes that we can ignore the diversity of individuals within a species and understand interactions simply on the basis of a typical or average individual. This is a good staring point for conveying principles that can be later elaborated with the added complexity of variety within a species.

In the case of a consumer-resource interaction, the arrow pointing from the consumer to the resource is denoted by a minus sign and the arrow pointing from the resource to the consumer is denoted by a plus sign (figure 2.1a) called a (+/−) link. This implies that the consumer derives a net nutritional benefit (hence +) by directly feeding on the resource; and the resource, being the victim suffers a cost (hence −). If the victim is another

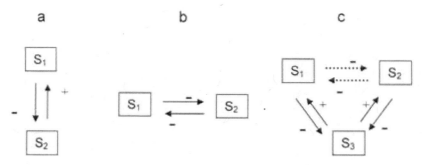

Figure 2.1. Three simple kinds of species interactions. (a) A direct consumer-resource interaction (solid arrows) in which species 1 (S_1) is the consumer and species 2 (S_2) is the resource. The consumer derives a net nutritional benefit (hence +) by directly feeding on the resource; and the resource, being the victim, suffers a cost (hence −). (b) A direct competitive interaction (soild arrows) between two species in which each species phys-ically preempts the other species' access to resources (hence a mutual −/− interaction). (c) A system in which two species (S_1 and S_2) vie for a common resource (S_3) through direct consumer-resource interactions. As a consequence, the two competitor species have a mutually indirect (hence dashed line) negative effect on each other mediated by the abundance of the resource species.

animal, then the cost is the victim's life. If the victim is a plant, then the cost is loss of some plant tissue such as leaves or stems. (Herbivores rarely kill and consume an entire plant—leaves, flowers, stems, and roots—in the same way that carnivores kill and consume their herbivore prey.)

If two species within a trophic level usurp one another's access to the same resources by holding territories or through direct physical struggles then each species pays the price for such interactions. Such competitors will have a direct mutually negative effect on each other's abundance. This is de-noted by two arrows each with a minus sign (figure 2.1b).

Two species may influence each other's abundance in less direct ways. Suppose that two species shared a common resource but never interacted directly with each other for access to that resource. In this case, the con-sumer-resource interaction reduces the availability of the resources through consumption. As a consequence, one species reduces the availability of re-sources for the other species. So, the one species has a negative effect on the other species. Here, both species again are competitors, but the effect is in-direct (as denoted by the dotted line) as opposed to direct (denoted by solid lines).

Thus, there are two kinds of competitive interactions. The first, in which species directly interfere with each other's ability to access resources, is known as contest or interference competition. The second form of competition is indirect and results from a mutual effort to exploit the greatest share of a common resource, called scramble or exploitative competition.

There are two kinds of competitive interactions. The first, in which species directly interfere with each other's ability to access resources, is known as contest or interference competition. The second form of competition is indirect and results from a mutual effort to exploit the greatest share of a common resource, called scramble or exploitative competition.

Conceptualizing Complexity: Direct Effects, Indirect Effects, and Species Diversity

The advantage of conceptualizing species interactions in terms of the couplets or triplets depicted in figure 2.1 is that we can assemble more complex structures by combining different couplets and triplets. For example, we could combine two consumer-resource couplets to create a linear, three-trophic-level food chain comprised of a top carnivore, a herbivore, and a plant species (figure 2.2a). We could build more branching systems by combining the basic exploitative competition triplet with two additional consumer-resource links to create a system in which two consumers each have exclusive access to a resource and share a common third resource (figure 2.2b). We could add a consumer link to an interference competition system such that a superior competitor is also more vulnerable to predation than a weaker competitor (figure 2.2c), called consumer-mediated competition. Alternatively, we could add two consumer links to two sets of interference competition systems (figure 2.2d) resulting in a trophically-mediated interference competition system.

The food web diagrams help to understand additional, important properties of species interactions in ecosystems. First, species diversity may arise as a consequence of many different dependencies or interactions. Second, whenever more than two species are linked together by direct consumer-resource or competitive interactions, we see the emergence of an indirect effect in which the middle species mediates the nature and strength of effect of the first species on the third species.

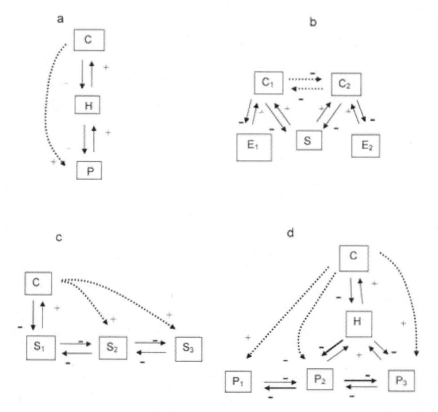

Figure 2.2. Complex food web structures assembled by linking combinations of the couplets or triplets presented in figure 5.1. (a) A three-level food chain in which a carnivore (C) directly feeds on an herbivore (H) that directly feeds on a plant (P). The carnivore has an indirect beneficial affect on the plant (dashed line) by virtue of suppressing the abundance of the herbivores eating plants. (b) A competitive system in which two consumer species (C) vie for a shared resource (S) and consume mutually exclusive (E) resources. (c) A system in which the abundance of a competitively dominant species (S_1) is controlled by a consumer species (C) thereby releasing less competitive species (S_2 and S_3) from domination. Consequently, the consumer has an indirect, beneficial affect on the less competitive species. (d) A multi-trophic variation of (c) in which a top predator (P) alters herbivore foraging and abundance on dominant and subordinate competitor species thereby introducing a host of indirect effects.

PREDATOR CONTROL OF SPECIES LOWER IN THE FOOD CHAIN

A classic indirect effect occurs in the three-trophic-level food chain (figure 2.2a) in which the top carnivore species (C) feeds on the herbivore species (H), which in turn feeds on a plant species (P). Because the carnivore reduces the abundance of herbivores that inflict damage to the plants, the carnivore provides an indirect benefit to the plants. That is, from the plant's perspective, carnivores fulfill the adage "the enemy of my enemy is my friend" (Holt 2001). Ecologists call this indirect effect a trophic cascade because the effects of manipulating the top trophic level of the system cascades down the food web to the lowest trophic level. This is the kind of indirect effect that is imagined whenever managers introduce predator species to control an outbreak of insect pests that are devastating to economically important crops.

EXPLOITATIVE COMPETITION—INGENIOUS WAYS OF DIVIDING-UP RESOURCES

A fundamental tenet of ecology is that no two competing species can co-exist by exploiting exactly the same resource—called the competitive exclusion principle. The species that wins in an exploitative competitive system (figure 2.1c) is the one that consumes the resource the quickest or draws it down to levels where individuals of the other species can no longer meet their food requirements and sustain themselves. But many ecological systems, especially ones that have species-rich plant assemblages, seem to contradict this fundamental tenet. Indeed, globally the approximately three hundred thousand species of terrestrial plants have only about twenty different limiting resources (light, water, CO_2, and minerals), and ecological science has shown that at most only three or four resources are limiting in any single location (McKane et al. 2002).

Resource limitation of plants is especially acute in arctic tundra systems in which plant growth is limited almost singularly by soil nitrogen availability. At the same time, many species of grasses, deciduous and evergreen shrubs, and forbs co-occur within areas of 0.1 square meter—a "paradox of diversity" (McKane et al. 2002). But, plants have evolved ingenious ways of dividing up the resources making the paradox only apparent.

Tundra plant species differ in rooting depth, timing of life-cycle development, and uptake preferences for different chemical forms of nitrogen (ammonium, nitrate, and amino acids). A sedge species (*Carex bigelowii*) uses mainly nitrate, which differs from a cottongrass species (*Eriophorum vaginatum*) and low-bush cranberry (*Vaccinium vitis-idaea*), which draw mainly from soil glycine and ammonium stores. Furthermore, low-bush cranberry obtains most of its nitrogen forms earlier in the growing season and at a shallower rooting depth than cottongrass. The herb Labrador tea (*Ledum palustre*) and the shrub dwarf birch (*Betula nana*) use mainly ammonium but differ in the seasonal timing of uptake. Each species divides up the nitrogen pool on the basis of chemical form, seasonal timing of use, and rooting depth. In other words, they minimize competitive overlap by dividing up the resource in ways that give each of them an exclusive spatial or temporal advantage.

> The species that wins in an exploitative competitive system (figure 2.1c) is the one that consumes the resource the quickest or draws it down to levels where individuals of the other species can no longer meet their food requirements and sustain themselves.

Many exploitative competitors also can co-occur when they share a common resource because they also have an exclusive resource (figure 2.2b). In such cases, no single competitor species can dominate the other because each species has a safety net in the form of a resource that is available for their own exclusive use. Thus, the shared resource does not determine the outcome of competition. Species diversity is maintained in this system because the consumer species (e.g., C_1 and C_2) mutually limit their abundance through exploitation of their shared resource species (S) and thus do not reach abundances in which they can overexploit their exclusive resource species (E_1 and E_2).

PREDATION: THE KEYSTONE OF ECOLOGICAL STRUCTURE

The examples presented above represent two ways that diversity originates and is maintained in ecological systems. In a food chain, species diversity is maintained vertically by consumptive interactions between trophic levels; in the competitive system, species diversity is maintained horizontally through fine partitioning of resources. In reality, many ecological systems are more complex than this because they are driven by combinations of vertical and horizontal diversifying mechanisms. We can begin to explore the kinds of complexity that arise by combining vertical and horizontal factors with two more examples that are effectively variations on a theme.

In figure 2.2c, the carnivore species again mediates competitive interaction, but in this case it is an interference competition system. Species diversity is maintained in this system because the abundance of a competitively dominant species, which would normally physically usurp space that could be occupied by other species, is suppressed by a carnivore. This allows less competitive species to gain a foothold in the system.

A textbook example of such an interaction occurs in rocky intertidal ecosystems where the starfish (*Pisaster ochraceus*) preferentially feeds on competitively superior barnacles and mussels. This allows other marine species including several species of algae, a sponge, and herbivores (limpets and chitons) to persist in the ecosystem. Removal of the starfish leads to a cascade of events including aggressive preemption of other species' space by barnacles. Barnacles are then overtaken by mussels. At the same time, many of the algae species are lost owing to a lack of space and their associated herbivores are lost due to a lack of food. A study of one such sequence of events found that diversity collapsed from fifteen to eight species (Paine 1966).

In a multitrophic level variation on this theme (figure 2.2d), a top carnivore (C) species interacts with an herbivore (H) species that consumes two or three plant (P) species leading to an indirect keystone effect. At the heart of this system lies an asymmetrical plant competition interaction in which the middle species, P_2, is competitively dominant to the other two species (depicted by a thick arrow of effect toward the other species). This can lead to a host of indirect effects. In one such field system (Schmitz 2003) a hunting spider carnivore (*Pisaurina mira*) interacts with a grasshopper herbivore (*Melanoplus femurrubru*), a grass species (*Poa pratensis*), and the competitively dominant species of goldenrod (*Solidago rugosa*). The grasshopper eats both the grass and the goldenrod. But, it prefers the grass in the absence of predators owing to its high nutritional value and can inflict considerable damage to it. Mortality risk caused by predator spiders causes grasshoppers largely to forego feeding on grass and to seek refuge in and forage on less nutritious but safer leafy goldenrod. This in turn causes high damage levels to this species. Thus, spiders exert strong cascading effects by having a positive indirect effect on grass abundance and a negative indirect effect on goldenrod abundance. Because of the spider's indirect effect on the competitively dominant plant, it releases other plant species from competition thereby having a positive indirect effect on plants not consumed by the grasshopper. The top predator here has an overall net diversity-enhancing effect on plants.

Both systems are examples of what is now called "keystone predation" (sensu Paine 1966): the former a direct keystone effect, the latter an indirect keystone effect. This concept derives its name from the keystone in an arch.

The structural integrity of an arch arises because an angled keystone at the top of the arch prevents the many (diverse) stones or blocks comprising the arch from falling into themselves and collapsing. Thus, if a predator that suppresses the abundance of a competitively dominant species is lost, species diversity in that system will collapse.

A limitation of using the food web diagrams to resolve complexity is that it assumes that these species interactions remain unchanged over time. Yet, species populations and hence their interactions can change over time because of one of the most fundamental processes driving life on earth: the evolutionary processes. In this process, diversity is effectively represented as different strategies that are pitted against each other in striving to achieve one goal: to maximize survival and reproduction.

If a predator that suppresses the abundance of a competitively dominant species is lost, species diversity in that system will collapse.

Life as a Game

The simple object of life's game is to contribute more genetic copies of yourself, offspring, to future generations than other members of your population (i.e., accrue as much natural capital as possible over your lifetime). Doing this successfully requires gaining the most resources possible and outwitting your enemies (competitors and predators). In this game, there are myriad strategies for success ranging from individuals pitting themselves against other individuals in outright competitive struggles to developing coalitions of cooperating individuals.

Natural populations are ensembles of individuals of the same species living together in some location. The strategies that individuals in these populations can use in the game are constrained by traits such as morphology, physiology, and behavior. No individual is exactly alike in any set of traits. Accordingly, no individ-

The simple object of life's game is to contribute more genetic copies of yourself, offspring, to future generations than other members of your population. Doing this successfully requires gaining the most resources possible and outwitting your enemies (competitors and predators). In this game, there are myriad strategies for success ranging from individuals pitting themselves against other individuals in outright competitive struggles to developing coalitions of cooperating individuals.

ual is likely to play the game exactly the same way. There are important implications that arise from such biological differences among individuals. In the extreme, any individual in a population that cannot outrun a hunting predator is doomed. So, slow-running individuals in a population may avoid death by hiding rather than running. There are, however, subtler ways that the game can be played.

Take, for example, the case of herring gulls caught during winter and placed on an ice block (Bartholomew 1977). During winter, they restrict blood flow to their legs and feet to minimize heat loss; thus, they can stand on the ice with no problem. However, these same birds increase blood flow to their legs and feet when experimentally exposed in an environmental chamber to warmer and warmer conditions that emulate the onset of summer. This is a physiological adjustment to increase heat loss. Now they melt into the ice when again placed on ice blocks. The seasonal adjustment in blood flow to the extremities is a physiological strategy or trait exhibited by all members of a herring gull population. However, individuals will vary in this trait. Most individuals likely will change their physiology quickly enough from winter to summer. Some, however, may acclimatize to the warmth very slowly because of a limited physiological capacity to adjust. The consequence is that as the environment becomes rapidly very warm, those individuals that acclimatize slowly become physiologically stressed.

Now, in the game of life organisms must deal with trade-offs because they have limited pools of resources or energy that can be allocated to the different competing demands. Individuals that allocate more energy to their own survival have less to allocate to offspring production, and vice versa. The consequence of this trade-off is that stressed individuals must allocate more energy toward their own survival (e.g., regulating their body temperature) than less stressed individuals: energy that could otherwise be allocated to reproduction. Chronically stressed individuals stand to survive less well and produce fewer offspring than less stressed individuals. They are said to be less fit than individuals with less stress.

Fitness

There is a special term applied to the quantity net survival and reproduction: it is called *Darwinian fitness*, *evolutionary fitness*, or simply *fitness* (Mayr 1982). The usage of the term here should not be confused with the vernacular that connotes physical health and well-being (e.g., aerobic capacity or endurance). This is not to say that physical health and well being do not have any bearing on fitness. Potential mates might find certain physical attributes attractive in the first place and perhaps be inclined to pair with healthy looking individuals more so than sickly looking individuals. But, physical health is only a proximate indicator of potential fitness. Fitness in a Darwinian evolutionary sense represents the lifetime contribution of an individual or a strategy to the breeding population, relative to other individuals or strategies. It is a measure of the relative success of a strategy in the game.

Individuals or strategies with low fitness (be it through lower offspring production or early death) fulfill the object of the game less well than individuals with higher fitness. Consequently, there will be fewer copies of themselves represented in future iterations of the game. That is, their strategy type or set of traits will become increasingly more rare or may even be absent in future populations.

> Fitness in a Darwinian evolutionary sense represents the lifetime contribution of an individual or a strategy to the breeding population, relative to other individuals or strategies. It is a measure of the relative success of a strategy in the game.

Essentially, what is being described here is a process that leads to change in the mean value of a trait or strategy in a population. It began with an ensemble of individuals that, on the average, have the capacity to make adjustments to environmental change modestly fast. As the environment changes, those individuals with difficulty coping become stressed and produce fewer genetic copies of themselves than individuals that cope well with the environment. The result is that the environmental change will favor certain coping strategies over others. The environmental gauntlet known as Nature determines whether or not a strategy stays in the game through a process of differential survival and reproduction. The process is known as natural selection. It is a process that begins with biological differences in some trait(s) among individuals and may lead to a change in the mean value of that trait in a population. The key point here is that *biological differences among individuals* exist. Without these differences natural selection could not happen and species populations will be unable to adapt to new environmental conditions.

Evolution by Natural Selection

Although natural selection sorts among biological differences in a population, such a process does not de facto mean that we should see a lasting change in the mean trait value or mean strategy in the population. That is, biological differences alone do not lead to the evolution of a trait or a strategy. The reason for this is that if no relationship existed between the trait value of a parent and those of their offspring, there would be no way that a parent could pass on the benefit of a particular trait to its offspring. So, the next reproductive period would produce offspring with the same degree of biological difference in traits as observed in the parental population.

Evolution via natural selection comes about when three conditions are fulfilled (Endler 1986). First, there must be biological differences among individuals in a trait that influences their capacity to cope with prevailing environmental conditions. Second, individuals with a particular trait must pass on that trait to their offspring. That is, offspring must inherit a particular trait from their parents. This will come about by passing a genetic code to offspring—called genetic transmission of a trait. Finally, there has to be a consistent relationship between a trait and fitness. Together, these three criteria can be used to explain how different traits or strategies come about in a population as a consequence of environmental change.

An example of this process is revealed in the evolution of beak depth among seed-eating finches (*Geospiza* species) of the Galapagos Islands. These are the celebrated group of finches that, in part, inspired Darwin to formulate his theory of evolution by natural selection. In this group of finches, beak thickness is important because it determines what kinds of seeds each species is adapted to utilize. Birds with thick, fat beaks are able to crush thick seeds that have a hard seed coat. Birds with narrower, nimble beaks are better suited to gathering and consuming thin seeds.

Gibbs and Grant (1987) conducted a long-term study during which time they made careful measurements on beak depths of birds within a population under different environmental conditions determined by a cyclic phenomenon of global weather change called El Niño. The measurements were made during El Niño years when there was plenty of rainfall and during drought conditions in years between El Niño events. Plants on the Galapagos produce seeds in all kinds of environmental conditions. However, the seeds of different plant species dominate in different conditions because different plant species do well in different environmental conditions. Drought conditions favor those plants that produce seeds with thick seed coats; thus there is a surplus of those seed types in the environment. Alternatively, rainy conditions favor those plants that produce thin seeds. In drought years then, there is abundant food for those individuals within a species that have comparatively thick beaks. Those individuals take in more energy than the thin-ner-beaked members of their population. Individuals with thicker beaks thus survive and reproduce better than individuals with thin beaks. Over time, the majority of individuals in the population will be thick beaked, which is reflected as an increase in the mean value of the measured beak depths among birds that remain alive in the population during and imme-diately after drought. Natural selection favors increasing beak depth. If the drought conditions persisted, or El Niño drought events become more fre-quent and longer lasting, we could eventually see the thin-beaked pheno-type become rarer and rarer and even disappear altogether.

The tendency to favor thick beak depth can be reversed when there is a period of rainy conditions and disproportionately higher production of thin seeds, because thick beaks are not sufficiently nimble to pick up these seeds. This results in a population containing individuals with comparatively thin-ner beak depths on average. So environmental conditions that fluctuate back and forth cause mean beak depths in a population to fluctuate back and forth accordingly. Traits that are favorable to individuals in one kind of en-vironment may not be as favorable in other environmental conditions. As the environmental conditions of the game—the playing field—change, so do successful strategies.

It is conceivable, then, that any human action that alters the environment, from local changes in land use patterns to serious insults such as habitat frag-mentation, pollution, and climate change on regional and global scales, has the potential to change the course of evolution. Human alteration of the environment imposes brand new nat-ural selection pressures on existing strategies within the world's biota. This can then kick off a string of changes in coping strategies. The im-plication here is that changes in indi-vidual strategies ultimately change the nature of the playing field and the game in a continuous feedback cycle.

> Traits that are favorable to individu-als in one kind of environment may not be as favorable in other environ-mental conditions. As the environ-mental conditions of the game— the playing field—change, so do successful strategies.

This is what ecologists call a complex adaptive system (Levin 1999). The perpetual feedback cycles that lead to adaptive change coupled with the myriad, simultaneous impacts that humans have even on a single location creates a high degree of uncertainty about the root cause of change in species composition and the functioning of ecosystems. This makes environmental problems "wicked" problems (Ludwig et al. 2001). In light of this, ecologists must be careful about the way they conduct their science to ensure that they gain reliable understanding of the way environmental change influences ecosystems.

Ecological Science: Gaining Reliable Knowledge about Ecosystems

Ecologists conduct their science by asking functional questions about organisms, their relationship to each other, and their relationship to the environment. Asking functional questions is a powerful way to study nature and contribute to environmental problem solving because it forces one to think about the root cause of a pattern or process. One way to begin deriving such a causal understanding of nature is to ask "Why" certain natural processes and patterns exist. Why questions, in a sense, are synonymous with functional questions because we must come up with answers that have an ultimate (cause-effect) basis, as opposed to a proximate (correlational) basis (Mayr 1982). To illustrate this point, consider the following example.

Suppose that you woke up one morning with a high fever. Fever is often a tell-tale sign that you have a disease. Technically, the fever is a disease symptom, a *proximate* indication that you have contracted a disease. But, many diseases can cause a fever—it is part of the body's normal immune response—so we don't know what kind of infection (e.g., bacterial, viral) caused that response. If the fever was severe enough, we might see a doctor to solve the problem with medication. The doctor has two choices. She or he could prescribe medication to reduce the severity of the fever. If this were the case, the doctor would be treating the symptom of the disease, or metaphorically, simply providing a *proximate* answer or solution to the problem. Prescribing medication such as antibiotics would be reasonable if the fever was caused by a bacterial infection. It would be an egregious mistake, if not malpractice, to give you the same prescription to treat viral meningitis or smallpox. Indeed society, through guidelines enforced by professional medical societies and the law, insists that doctors do not merely treat disease symptoms, but rather

Asking functional questions is a powerful way to study nature and contribute to environmental problem solving because it forces one to think about the root cause of a pattern or process.

give a proper diagnosis for the cause of the disease. In other words, doctors are expected to understand the cause to give the correct solution to the problem.

The problem facing ecologists, however, becomes more complex when the patient is the entire natural system. Take, for example, the case of Lyme disease. It is a serious human ailment especially in the northeastern United States (Barbour and Fish 1993). It causes debilitating arthritic conditions and even serious nervous disorders in infected individuals if the disease is not diagnosed and treated early on in the infection stage. Lyme disease is transmitted to humans from a species of tick. These ticks normally feed on the blood of wildlife species such as white-tailed deer (*Odocoileus virginianus*) and deer mice (*Peromyscus sp.*), but they will feed on humans given the opportunity. Humans contract Lyme disease locally when the disease-causing spirochete passes from the tick into the human bloodstream while the tick is obtaining a human blood meal. Immediate symptoms of the disease include a bulls-eye target rash surrounding the location where the tick obtained its blood meal, followed by a high fever. One solution is to prescribe antibiotics that kill the spirochete after verifying that the patient has contracted the disease. From a medical standpoint, this is treating the root cause of the *human* disease condition. From an ecological standpoint, however, it is merely treating the symptom. Antibiotics will never eradicate Lyme disease.

The way to begin controlling Lyme disease on a human or landscape scale is to understand why ticks, deer, and deer mice populations thrive together in the semi-rural environment of northeastern United States in the first place; and why they sustain spirochete populations. This is what ecologists try to do. However, environmental problems like this are very complex because they are often intertwined with a variety of other environmental factors and they usually occur on an extraordinarily large scale. Furthermore, they are often consequent to long-term changes humans have caused to natural landscapes. In the case of Lyme disease, this involves interdependencies that carry over many years. For example, the effect of mass acorn production in a given year is only realized two years later by increased abundances of mice that in turn are hosts for juvenile ticks that in turn carry Lyme disease. The increase in host abundance means that the ticks have a higher chance of obtaining the blood meals needed for survival, leading to increased abundance of the carriers of the disease. It took thirteen years of field research to identify this causal chain of interdependency and rule out other potential explanations (Ostfeld et al. 2006).

Environmental problems are thus typically more difficult than medical problems to solve because the chain of causal events is more difficult to trace. Sorting through such complexity scientifically is not insurmountable, but it represents the single most important challenge facing ecological science (Levin 1999). It requires following a systematic procedure that allows one to develop understanding of how the components of natural systems fit together and function. That systematic process is scientific methodology.

In attempting to provide reliable scientific knowledge, ecology, like any other scientific discipline, follows certain scientific methodology. Ecologists use any or all of several scientific methods including (Romesburg 1981; Mayr 1982):

1. Induction
2. Retroduction
3. Hypothetico-Deduction

These methods offer different perspectives about nature and the degree of causal understanding of its functioning. In the following, I provide a brief overview of these methodologies.

Induction

Induction is the fundamental step in any scientific enterprise. Induction provides the foundational understanding of natural pattern or process that demands explanation. In an ecological context, induction is the formalization of natural history observations. Let me illustrate with an example.

An important goal in ecology is to derive explanations for patterns of species diversity in nature. Suppose we asked the question: What is the relationship between the diversity of insect species in an area and the diversity of herbaceous plant species comprising their habitat? Suppose also that there were no published data yet available to answer this question. The only recourse, then, is to go into the field and gather those data by sampling.

Let's assume that we chose to sample in three different areas. The three areas range from a plant monoculture in the low diversity case to a multiple species plant mix in the high diversity case. Let us assume, for the sake of argument, that species in our example can be identified by their appearance, or morphology. A simple count of morphologically different types reveals that the low diversity area has four insect species residing on the single plant species. A similar count reveals that the medium diversity area contains two plant species and six insect species and the high diversity area has three plant species and eight insect species. An answer to our initial question is that insect species diversity increases with plant species diversity. That is, there appears to be a positive relationship between plant and insect species diversity.

Induction here led to the discovery of an association between two kinds of observations or variables (known as correlation). In our particular example, a positive correlation implies that plant species diversity is somehow related to insect diversity. There is, however, one important hitch. Although we have identified a relationship, we still haven't explained *why* higher plant species diversity is associated with higher insect species diversity. Formulating explanations for observed phenomena lies within the next phase of science methodology.

Retroduction

Retroduction ascribes a reason—or technically an *hypothesis*, because the reason is only proposed and thus not yet validated by a scientific test—for the association or trend observed through induction. This is perhaps the most creative aspect of science. The scientist uses his or her imagination in combination with accumulated knowledge about nature to explain why we see a particular phenomenon or trend. This phase of science often leads to those new ideas or theories that form the important conceptual foundation for a discipline.

In our example, one plausible explanation for the positive correlation between insect species diversity and plant species diversity is this: A greater variety of plant species provides a greater variety of resources for insect species with different food requirements. Therefore habitats rich in plant species offer a host of food resources to support a rich variety of insect species.

This may seem like a satisfactory answer. Nonetheless, it remains untested. So, we don't know if our particular explanation is the correct one. Without testing the hypothesis, the knowledge we acquire from this stage can be shaky. One must be mindful that all hypotheses, however well conceived, could be flatly wrong. Serious negative consequences can ensue if we stopped at this stage and applied this knowledge to policy and management (Romesburg 1981). The danger here is that we may create significant changes to ecosystems before we really know whether or not the hypothesized causal relationship exists in nature. Hypotheses must be tested before applying them widely to environmental problem solving because equally plausible alternative hypotheses to explain a particular pattern or process may exist.

To know if a hypothesis is a reliable explanation, it must be tested experimentally by manipulating the putative causal variable(s). This is the domain of the Hypothetico-deductive method.

Hypothetico-deduction

In the Hypothetico-deductive (HD) method, one deduces testable predictions based on the hypothesis. Then, most critically, one tests the hypothesis.

Let me illustrate by building on our example. Suppose that we chose to test the hypothesis that higher plant species diversity leads to higher insect diversity because of the inherent food value afforded by the variety of plant species. One possible food source for many insects is flowers because they provide nectar and pollen. Each plant species will have unique flower characteristics that attract different insect species. So, more plant species lead to a greater diversity of flower types, which then offers a greater variety of pollen and nectar resources. The logical deduction then is that if we manipulated the diversity of flowers in a patch, we should see an associated change in insect species diversity.

One way to experiment with flower diversity is simply to cut off selected flowers from plots in the field containing high plant diversity. Such a manipulation is called an experimental treatment or experimental perturbation. If our hypothesis (and associated logical deduction based on it) is correct, then we should see a decline in insect species richness in the plot following the application of the experimental treatment. If we see the decline, we may conclude that a higher diversity of food leads to higher insect species diversity. But, is this conclusion reliable? The answer is no. This is because under the current experimental design we have failed to include an experimental control.

The experimental control serves as a critical, unmanipulated baseline for comparison with the experimental treatment. It is established simply by leaving some plots alone, or technically leaving plots "untreated." The control allows us to tease apart the effects of the experimental treatment from random and potentially confounding environmental effects. For

The point is if we do not have an experimental control, then we cannot draw reliable conclusions about our hypothesis test. No Control = No Conclusion.

example, suppose that during experimentation we encountered a period of unusually cold days that caused many insects in the field to crawl under leaves on the ground and take advantage of the insulative value of the leaf litter. Such a temperature change also could lead to a decline in the number of insect species observed in our experimental plots. If temperature change was the driver of insect abundance, then we should see a decline in both treatment and control plots. So, without a control, there is a risk that we may falsely conclude that the experimental manipulation of food caused the decline in insect species diversity. The point is if we do not have an experimental control, then we cannot draw reliable conclusions about our hypothesis test. No Control = No Conclusion.

The discussion of ecological complexity, evolution by natural selection and science methodology provided above has equipped us with working foundational concepts. These concepts are variously applied in the rest of the book to show how ecological science can offer a scientific route to understanding ecosystem structure and function, and to offer policy makers and managers insights to ensure the sustainability of ecosystems.

3

Climate: Template for Global Biodiversity

THE PHYSICAL ENVIRONMENT, THE NONLIVING PART OF OUR WORLD, SETS the background for all living beings. A major factor of the physical environment is climate, which ultimately determines water availability and thermal conditions. These two factors interact to determine how an amazing variety of organisms are distributed in different parts of the world. But what determines climate? Climate is determined by interactions between the sun, as producer and emitter of energy, and the earth as both a receiver and transmitter of energy. The nature of the energy exchange between sun and earth is what eventually leads to patterns in the distribution of life on earth.

The Physics Underlying Life on Earth

The sun radiates energy into space. That energy is absorbed by any body that is cooler than the sun that has mass. One such body is the planet earth. Because of its proximity to the sun, solar radiation strikes the earth at very high intensity. That energy is absorbed at the earth's surface, and because the earth has mass, there is a high storage capacity for that energy causing the planet to heat up. If all the energy striking the earth were just stored, however, the planet would reach exceedingly high temperatures and then vaporize. The reason that this doesn't happen is that the earth, like all absorbing bodies, reradiates much of that heat back into space. Eventually, absorbed energy is exactly balanced by reradiated energy, according to physical laws, leading to a steady state heat budget. The specific temperature, at steady state, depends upon the make-up of the absorbing body (e.g., mineral rock, water, woody tissue, etc.). The estimated average temperature for the planet, which is predominately rock and water, is $-21°$ C. But, a temperature of $-21°$ C means that the earth should largely be a frozen ball of ice, not the green-blue planet we see in satellite images from outer space.

The factor that we have neglected to consider in our calculation is the earth's atmosphere. Or more directly, we have neglected to consider the effect of certain gases—water vapor (H_2o), ozone (O_3), methane (CH_4), and carbon dioxide (CO_2)—in the earth's atmosphere. These gases together make the atmosphere opaque. This opacity does two things. First, it prevents a good amount of (but not all) the solar radiation from striking the planet by reflecting that radiation back to outer space, which in turn reduces the amount of potentially harmful radiation like ultraviolet rays from striking the earth. This is why we worry about holes in the ozone layer of the atmosphere.

Nevertheless, some radiation has wavelengths small enough to pass between the gas molecules and reach the earth's surface. As the earth's surface heats up, it becomes much warmer relative to the surrounding air and so reradiates heat energy back toward outer space. The energy that is emitted is now in the form of a longer-wave radiation called infrared radiation that has difficulty penetrating through the layer created by the atmospheric gases. This energy would be completely lost were it not for the absorbing capacity of the gases. The energy absorbed by the gases is then reradiated back to the earth resulting in a moderation of the earth's climate. We have all experienced evidence of this moderating effect during the course of our daily lives. For example, summer nights with clear skies often require one to don a sweater because the temperatures are much colder than summer nights with opaque (cloudy) skies, even though daytime temperatures could have been identical.

We apply this basic principle of physics whenever we warn society not to leave their pets in cars with closed windows on hot summer days or when we construct greenhouses for growing plants. Consequently, the energy absorbing and moderating effect of these gases on the planet is metaphorically called the *greenhouse effect*. The gases that contribute to this effect are accordingly called *greenhouse gases*. The greenhouse effect caused by the presence of these gases is what leads to an estimated steady state temperature of $15°$ C for the planet.

Consequently, the energy absorbing and moderating effect of these gases on the planet is metaphorically called the *greenhouse effect*. The gases that contribute to this effect are accordingly called *greenhouse gases*.

Now, we know that there are places on earth such as arctic regions that rarely if ever reach 15° C and other places such as equatorial tropical regions and deserts that routinely exceed this temperature. What scientific value, then, is there to an estimate of 15° C? There are two values. First, this temperature represents a baseline *average* for the entire planet. This average accounts for regional temperatures that may fall below or exceed this value. Second, the recognition that this is an average and not applicable to the entire globe generates questions about why there are regional differences in temperature to begin with. The answer lies again in considering the nature of the sun as producer and emitter of radiant energy and the earth as absorber of that energy.

Like the sun, the earth is a ball. The earth is, however, far enough from the sun that all beams of the sun's radiation can be considered to be parallel to each other. The combination of parallel incoming radiation and the curvature of the earth results in differential intensities of heat radiation that strike the earth's surface at different locations. If we measured the amount of solar radiation striking a fixed-size plot near the equator and at the pole, we find that the intensity of radiation striking the plot near the equator is far higher than that striking the same-size plot near the pole. That is to say, the intensity of radiation striking a unit area is far higher at the equator than at the pole (figure 3.1). As a consequence, there is less energy to heat up a unit area at the pole than at the equator. Less energy absorbed means lower realized temperatures.

The differential thermal regimes when coupled with the earth's rotation then generate prevailing winds and ocean currents. Warm water or air flows to colder regions; colder water or air circulates back to warmer regions. This circulation pattern, coupled with the earth's geographic formations (e.g., mountains, flatland, etc.), determines the kind and amount of precipitation that occurs at different geographic locations.

Ecosystem Types

The combination of temperature and moisture determine patterns in the distribution of life on earth. A method to quantify pattern in global life zones or ecosystem types using such climatic data was developed by Holdridge (1947). These ecosystem types range from tundra and taiga in polar and alpine regions, temperate deciduous and coniferous forests in the northern and southern latitudes, tropical rainforest, savanna grasslands, and deserts in the midlatitudes. Figure 3.2 demonstrates that different regions

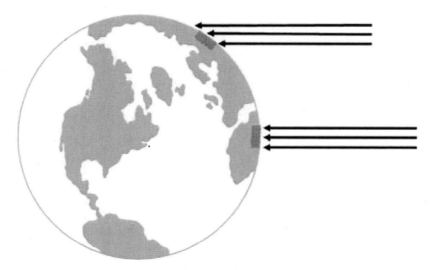

Figure 3.1. Illustration of differential heating of the earth's surface leading to global differences in temperature. The arrows depict parallel rays of sunlight striking the earth's surface. Due to the earth's curvature, the intensity of solar radiation (depicted as the number of arrows) striking a unit area of surface (grey rectangle) is less in the polar regions than near the equator. Lower intensity of radiation at the poles translates into less heating of the surface than near the equator.

share physical attributes even though they are comprised of different kinds of species. For example, arctic tundra regions are effectively deserts with cold temperatures; and rainforests are not found solely within warm tropical regions.

Coping with Climate

A species' capacity to cope with its surrounding physical environment is determined by its physiology. Physiological processes operate at different rates under different conditions. For example, rates of photosynthesis and respiration (burning of food energy) are temperature related. Other processes depend upon water or nutrient availability. A species' performance in or tolerance of local climatic conditions is defined by certain limits.

Tolerance can change over the short term (individuals' lifetimes) as individuals become exposed to seasonal changes in environmental conditions. For example, with the onset of winter warm-blooded species such as white-tailed deer (*Odocoileus virginianus*) face cold temperatures and poor quality

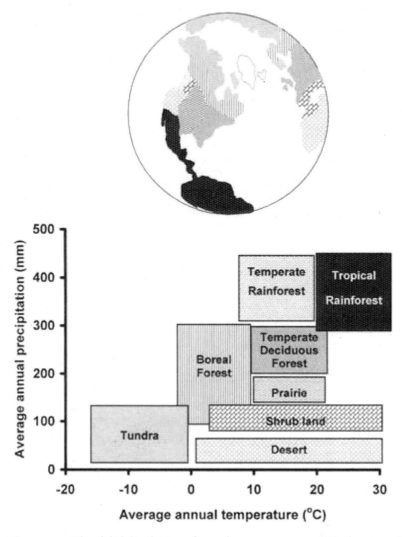

Figure 3.2. The global distribution of some key ecosystem types. The location of specific ecosystems is determined both by average annual temperature and by average annual precipitation. The figure shows that deserts are not strictly in hot regions: arctic tundra is effectively a cold desert. Likewise, rainforests are not restricted to tropical regions.

forage (woody twigs). The rate of heat loss to the environment is propor-
tional to the difference in surface temperature of a body and the temperature of the surrounding environment. To lower the rate of heat loss, deer
increase the thickness of their undercoat to decrease the amount of warm,
bare skin exposed to the cold elements. In addition, the thin legs of deer
have a high surface area to body volume ratio. This is akin to having large
picture windows in a tiny bungalow. Essentially, considerable amounts of
heat within the core of the house (or the deer's leg) will radiate out to the
environment because there are many large yet poorly insulated surfaces over
which heat is quickly exchanged with the environment. To reduce heat loss
deer restrict the amount of warm blood flowing to these extremities and
they lay down fat to insulate the extremities with body tissue that is not
highly prone to freezing. Deer also compensate for poor quality forage by
lowering their metabolic rate during winter to decrease their demand for
food energy. These changes are reversed with the onset of warmer temperatures and better quality forage in spring.

Climatic Conditions and Performance

Different species are located in different geographic regions in part because
they have different capacities to deal with regional differences in climatic
conditions. For example, consider two species that have different abilities to
tolerate temperature and direct solar radiation. Differences in temperature
tolerance may arise because some species such as birds and mammals have
feathers or fur that insulate their body whereas other species such as reptiles
and amphibians do not. Differences in solar radiation tolerance may arise
because of differences in the color of the body surface, which determines
the amount of solar radiation absorbed: darker surfaces absorb more solar
energy than do lighter surfaces. We can depict these species-specific abilities to tolerate temperature and solar radiation by plotting individual fitness
(that is individual survival and reproduction) against temperature and against
absorbed solar radiation (figure 3.3). These curves can be determined experimentally by placing individuals of a species in different temperature
regimes, while holding the amount of incoming solar radiation constant,
and by placing individuals in different solar radiation regimes while holding temperature constant. The net result is that we tend to find that individuals of a species have a temperature or solar radiation level where they
exhibit peak performance (high fitness) and there are neighboring temperatures or radiation regimes where they can perhaps do all right (medium to

low fitness). Finally, there are extreme temperatures and radiation regimes that these species cannot cope with and so exhibit zero fitness. This is because extreme temperatures or solar radiation levels could cause species to overheat or freeze to death depending on their physical (e.g., presence of fur or feathers) and physiological (e.g., ability to adjust metabolism) traits. Note that figure 3.3 also depicts an important characteristic of species, namely a trade-off in coping abilities. Species 2 is able to tolerate higher temperatures than species 1 but it is less able to tolerate higher solar radiation regimes than species 1. These trade-offs are noteworthy because they indicate that species cannot be good at coping with everything.

The net result is that we tend to find that individuals of a species have a temperature or solar radiation level where they exhibit peak performance (high fitness) and there are neighboring temperatures or radiation regimes where they can perhaps do all right (medium to low fitness).

The species-specific information provided in the upper graphs in figure 3.3 can be combined to create a "contour map" depicting different levels of fitness over the entire range of temperature and solar radiation that each species has been determined to tolerate. The value of this information is that it can help us understand how individuals of a species might play the game of life. For example, individuals of species 2 will have the highest fitness in regions with high temperatures but modest solar radiation. These conditions are considered optimal for that species. Individuals of that species are then expected to seek out habitats on a landscape that provide such optimal conditions for living. Environmental conditions could change such that the optima shift to a new location. For example, species 2 might be associated with meadow habitats. As plant succession fills in the meadows with trees, and it thus becomes cooler with less solar radiation hitting the ground, the location is no longer optimal for this species. Individuals will either have lower fitness or even die if they remain at this location. Or they can move and seek out new meadow habitats on the landscape that offer optimal conditions. The ability to move and seek optimal conditions obviously varies with the kind of species. Individual trees that are firmly rooted in a location have no recourse to move. Animal species have considerable flexibility to move and seek better conditions.

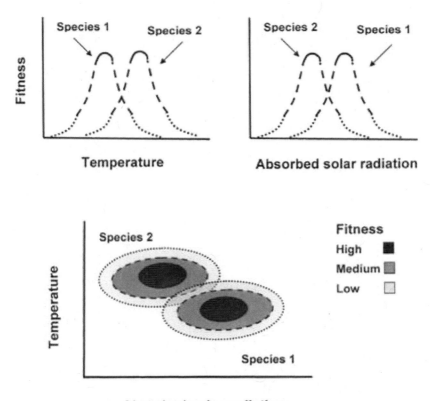

Figure 3.3. Hypothetical example of temperature and solar radiation effects on fitness of individuals of a species. The composite figure presents fitness contours that represent the range of temperature and solar radiation under which individuals in a species can thrive at optimal levels (black zone) to conditions in which they have a modest level of survival and reproduction (medium gray to light gray zones). The light gray bounded by the dotted line represents conditions likely found at the geographic range margins for the species. The figure illustrates that species 1 is better able to cope with lower temperatures but higher solar radiation than species 2.

Climatic Conditions and Adaptation

From a functional standpoint, we typically want to know why a particular species came to cope with a certain set of conditions in the first place. Understanding why species have these abilities allows us to make some reasonable forecasts about what impact changing climate might have on these species in the long run.

Whenever we want to derive an ultimate answer to a functional question we need to appeal to an evolutionary process. The one used in this book is the principle of adaptive evolution by natural selection. Let's now apply it to see how different species might have come to tolerate certain environmental conditions.

We begin with a rabbit species that lives in semi-desert shrub land ecosystem at middle to southern latitudes. Hot temperatures and high levels of direct solar radiation typically characterize this life zone. Let's assume that individuals of this species live there because this environment offers optimal climatic conditions for that species. Individuals of this rabbit species tend to be very large (about the size of a beagle dog); they have long ears and long, gangly legs and a very thin fur coat.

Suppose that the region providing optimal conditions is crowded so that individuals residing there compete fairly intensely for resources that are important to fitness (e.g., vegetation). In such a case, some individuals (less competitive ones) may be preempted from getting enough resources to support their survival needs. Suppose that these individuals are forced to relocate to more northerly, cooler conditions—otherwise they would die. In the new locations, they live under less than optimal conditions. These individuals may survive and reproduce at these new locations, but not to the same extent as their counterparts that live under optimal conditions. Indeed, individuals that live under extreme temperature and solar radiation conditions that push their tolerance limits may not survive at all. One reason that those individuals might not live is that their large bodies and thin coats result in a high surface area to body volume ratio with limited insulation. In addition, traits—long ears and gangly legs—that allow this species to cope well in hot conditions by rapidly dissipating heat now become a penalty because they exacerbate heat loss in an environment where it would be better to conserve heat.

Now, individuals in a population do not have identical body structure. There is variety in populations. Suppose that, as part of the variety, there were individuals living at the cooler margin that had a somewhat smaller body size and smaller ears and legs and a slightly thicker fur coat. These individuals would not lose body heat as rapidly as their larger, gangly counterparts. Essentially, we now have two strategies in the same population vying for existence at the cooler margin. However, individuals with the smaller body plan have a distinct advantage because they are able to conserve heat better than larger individuals. Suppose that this energy saving can be allocated to reproduction and that small parents tend to produce off-spring with a similar body plan. Individuals with the smaller body plan, a smaller *phenotype*, will then have higher fitness at the cooler margin than their larger counterparts and over time come to dominate the population at the margin. Suppose now that small individuals preferred to breed with similar-size individuals (what evolutionary biologists call *assortative mating*, [Mayr 1982]) and that larger individuals tended to avoid the cooler margin whenever possible. The consequence is that eventually, smaller individuals might only associate with other smaller individuals and thus create populations of their own that eventually do not interbreed with larger individuals. Such reproductive isolation, over a long period of time, may eventually lead to a new species. That is, adaptation via natural selection has led to a new species that tolerates the cooler conditions much better than the large individuals of the other species from which it originated.

Climate-Space

One of the difficulties in forecasting the effects of climate on species is that it is hard, in practice, to measure individual fitness under field conditions. As a consequence, ecologists (e.g., Porter and Gates 1969; Gates 1980) have used principles of energy physics to define the larger envelope of climatic conditions that a species could tolerate. This now classic approach is called a *climate-space* analysis (see Box 3.1).

Organisms tolerate temperature changes through day-to-day physiological and behavioral adjustments and seasonal acclimatization. But, coping with climate isn't simply about coping with temperature. Organisms also must tolerate changes in incoming solar radiation and radiative energy exchanges with their surrounding environment. To illustrate this principle consider an elk (*Cervus elaphus*) lying on a patch of grass on a hillside covered by trees and rocks.

The elk will lose heat through several processes. It will radiate heat to the environment if its body temperature is warmer than the environment. It will conduct body heat to the ground upon which it is lying. Wind will draw heat away from the body surface exposed to the air through convection. The stronger the wind, the faster heat will be lost. It will lose heat by panting. The moisture in the mouth leads to the same kind of evaporative cooling as in humans when they sweat. Unlike humans, however, elk and many other species pant because they do not have sweat glands.

The elk will also gain heat from several sources. The elk will absorb incoming solar radiation. Like the elk, the surrounding rocks and trees absorb solar radiation and reradiate that energy back to the environment, some of which is absorbed by the elk. The elk's own metabolism will produce heat much like a furnace in a house. The elk must balance the exchanges of heat energy with the environment in order to prevent overheating or freezing. That is, the elk reaches its own steady state body temperature by balancing absorbed and emitted radiation. The elk does this through behavioral and physiological means that are determined by its phenotype. Thus, the physiological and behavioral traits that make up its phenotype determine the range of environmental conditions that it can tolerate, as illustrated with the rabbit example above. The climate-space approach allows us to formalize these physical processes and quantitatively account for the effects of the thermal environment on an organism's energy (heat) budget. The climate space effectively represents the outer boundary of a fitness "contour map." The power of a climate-space approach is that it can be extended to understand and forecast the effects of climate change on a species' geographic range, as is demonstrated in Box 3.1.

Effects of Global Climate Change

We now are entering an era where global climate change, brought about by rising levels of greenhouse gases, has the potential to alter the distribution of life on the planet as we know it. It is easy to deduce using the principles described at the beginning of this chapter that higher atmospheric concentrations of greenhouse gases will lead to rising average global temperature. But, what exactly is that new temperature? What will be the attendant consequences of that temperature rise on the diversity of life on earth?

Box 3.1 Climate-Space Analysis

A climate-space analysis is an approach that quantifies how solar radiation, environmental temperature, and physical and physiological processes (conduction, convection, evaporative cooling) influence the exchange of heat energy between an individual organism and its environment. A climate-space delineates the range of environmental conditions that an individual species can tolerate and still survive.

The approach generates a climate envelope on a graph (see below) relating ambient air temperature and absorbed solar radiation (Qa) by species. Qa is determined by characteristics of a species such as surface color, thickness of fur, body shape, and thickness of body fat in conjunction with the level of direct and indirect solar radiation striking the body surface of the species. In the envelope, lines with positive slope represent combinations of minimum nighttime (top left line) and maximum daytime (bottom right line) temperature and solar radiation that can be tolerated. The lines on the left and right sides of the envelope are combinations of temperature and solar radiation that can be tolerated when the animal has attained the minimum (left line) and maximum (right line) allowable body temperatures. The animal incurs injurious, if not lethal, effects when body temperatures exceed these limits. The climate-space approach has been validated with numerous small-scale experiments and field observations (see Gates 1980).

The climate-space approach can be tested at the geographic scale of species distributions. The figure below depicts the range distribution (light gray area) of a hypothetical species that lives in southwestern Canada. If the climate-space gave correct insight, then measures of ambient temperature and absorbed solar radiation at explicit locations within a species current range distribution (light gray circles) should fall within the climate envelope. Values measured outside the geographic range (dark gray circles) should fall outside the envelope. This has indeed been shown to be the case (Johnston and Schmitz 1997).

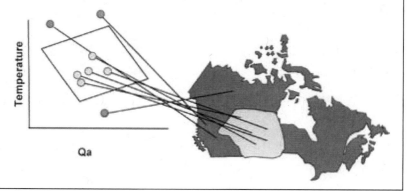

Box 3.1   *Continued*

Once the correspondence between climate-space and current species distribution has been established, it is a straightforward exercise to forecast where species may relocate geographically (i.e., redraw the range boundary) if climate change caused temperatures to increase or decrease on a geographic scale. Essentially one compares the new geographically explicit distribution of temperature and solar radiation with the climate space envelope and discerns which new climatic conditions can or cannot be tolerated.

The environmental science community has worked feverishly for most of the last two decades to answer just that question (IPCC 2001). The current expectation is that there will be at least doubling of mid-1980s level CO_2 emissions by the year 2050. The effect of this on future climate has been forecast using General Circulation Models (GCMs). GCMs are designed to provide, on a global-scale grid (i.e., a 0.5° Latitude/Longitude grid), information such as daily and monthly maximum surface air temperature, monthly precipitation, total incident solar radiation, and surface wind speeds. These models are first calibrated by comparing simulated base climatic conditions (current CO_2 levels) with existing conditions. They are then used to forecast the plausible effects on climate of a doubling of atmospheric CO_2. Each of the models paints a slightly different picture of change, depending on their built-in assumptions. Collectively, however, they are consistent in their prognosis for extent and range of effect, namely a rise in mean temperature between 3° C to 5° C , thereby increasing our confidence that doubling atmospheric CO_2 is a phenomenon that has a high likelihood of occurring. Such changes have occurred throughout the geological history of the earth. The difference between then and now is the rate at which this change is taking place. For example, current projections show that mean global temperature will rise 3 to 5 degrees within the next fifty to one hundred

Such changes have occurred throughout the geological history of the earth. The difference between then and now is the rate at which this change is taking place. For example, current projections show that mean global temperature will rise 3 to 5 degrees within the next fifty to one hundred years—a change that required thousands of years in the geological past.

years—a change that required thousands of years in the geological past. One might argue that this, nevertheless, is a minor change. But remember, this is 3 to 5 degrees above a 15 degree Celsius global mean temperature—or a 12 to 30 percent change.

The key issue in conducting assessments of climate change on species is that it cannot be done using the long-standing, time-tested tradition of conducting hypothetico-deductive research in which one carries out detailed experiments to test hypotheses that lead to cause-effect insights. This naturally creates a dilemma for environmental policy because one cannot obtain the kind of strong empirical evidence normally needed to effect policy change. One way to circumvent this limitation is to develop forecasting tools (models) that build upon the physiological models that have been calibrated empirically via small scale, replicated experiments and link them with large-scale output from climate models. This approach links the ecology of organisms at smaller scales with climate change data that are relevant at much larger scales (Pacala and Hurtt 1993). We can then examine, through computer simulations, the climate sensitivity of a species (i.e., how much a species' geographic range distribution changes with a change in climate).

Forecasting Effects on Wildlife Species

Such an exercise was undertaken to evaluate the sensitivity to climate warming on several mammal species (elk, white-tailed deer, Columbian ground squirrel [*Spermophilus columbianus*], and eastern chipmunk [*Tamias striatus*) within the continental United States (Johnston and Schmitz 1997). These species were chosen for several reasons. First, they represent the kind of species that immediately come to mind when society thinks about biodiversity. Second, the ability to cope with heat stress varies with body size, and the species chosen are near the endpoints of the spectrum of body sizes of North American mammals. Finally, climate change effects will differ between the western and eastern United States. The first two species have largely eastern distributions, and the latter two have western distributions.

Climate change is expected to affect wildlife species in two ways. First, it could *directly* affect an animal species by compromising its ability to cope with anticipated levels of heat. This arises because excessive heat can have injurious effects on biochemical and physiological processes especially in locations where species already are living close to lethal temperature limits. Second, climate change is likely to impact wildlife species *indirectly* by causing sweeping changes in continental scale distributions of their habitat.

DIRECT EFFECTS

Following the approach outlined in Box 3.1, the direct effects of climate on wildlife species was evaluated by gathering data on solar radiation, environmental temperature, physical (conduction, convection), and physiological processes (evaporative cooling) for different locations within and outside of each species known geographic range. These values were then compared against the climate envelope defined for each species by a climate-space model. An example climate-space diagram for current (1990s) temperature and incoming solar radiation is presented for elk in figure 3.4. The data values lie in the midrange of the climate envelope, which is to be expected if each species were living at or near its climate optimum. Figure 3.4 also presents the expected combinations of air temperatures and solar radiation exchanges that elk are expected to encounter under a doubling of atmospheric CO_2 in the hottest part of the year (July climate) within their current geographic ranges. All of the data points still fall within the climate envelope, indicating that these particular species should have the physiological capacity to tolerate anticipated levels of climate warming. Note, however, that the data cluster has moved toward the edge of the climate envelope, relative to 1990s, indicating that conditions will move away from the optimum. This implies that elk may see a decline in fitness at their current locations and so may have to migrate to new geographic locations that offer optimal conditions. These conclusions extend also to the other three mammal species studied.

INDIRECT EFFECTS

It is likely that the impact of global warming extends far beyond the direct effects it has on wildlife species. Indirect effects including response of wildlife to shifting habitat are also likely to be of significant importance. Thus, a more complex analysis is required to examine the effects of distributional changes in vegetation communities that comprise habitat on the distribution of wildlife species. Such an analysis requires obtaining forecasts of climate change effects on the geographic distribution of vegetation communities and relating that to the distribution of wildlife species (e.g., Johnston and Schmitz 1997).

Geographically explicit information on climate change effects on plant communities has been generated by the VEMAP project (VEMAP members 1995). This project developed very detailed models that simulated the physical processes determining the biotic composition of vegetation communities. These models were then linked with GCM forecasts for current

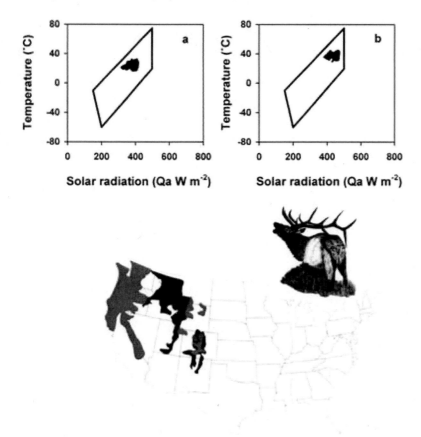

Figure 3.4. Climate-space diagram showing temperature and solar radiation values for coordinates within the current range distribution of elk under (a) current 1990s conditions and (b) under a doubling of atmospheric CO_2. Injurious or lethal effects would only occur if solar radiation or temperatures fell outside the limits. Note that all the data points lay within the thermal limit or climate envelope, implying that elk can tolerate the direct effects of climate change. The figure also presents the indirect effects that climate change could have on the geographic range distribution of elk within the United States. The black regions represent locations where elk currently reside and where they are expected to reside under climate warming. The gray regions represent locations where elk currently reside but where they could disappear under climate warming due to loss of suitable habitat. All figures drawn using data presented in Johnston and Schmitz (1997).

and future climate. This modeling predicts some significant northward shifts in major vegetation communities within the continental United States. For example, the eastern United States is expected to lose much of its cool-temperate mixed forest because it is expected to shift northward into Canada. This will be replaced by expansion of a warm temperate mixed forest that is currently characteristic of southeastern United States. Prairie habitats that characterize north-central states will be replaced by range expansion of prairie habitat that is characteristic of south-central states. Temperate arid shrub lands, alpine tundra, and taiga in the intermountain region will disappear from most of their current range.

The effect of these vegetation shifts on wildlife species distributions was evaluated by following several steps. First, maps of current species distributions were overlaid onto a map of the current continental United States distribution of ecosystem types and the statistical relationship between them was estimated. This statistical estimator was used in conjunction with a new map, depicting ecosystem change in the face of climate warming, to assess climate impacts on the two wildlife species. In general, white-tailed deer, eastern chipmunks, and Columbian ground squirrels were expected to retain the same range size or increase it. The geographic distribution was expected to shift somewhat for white-tailed deer, almost entirely for Columbian ground squirrels, and not at all for eastern chipmunks. Elk were expected to suffer the greatest impact with a 93 percent loss of current range and no prospect for range expansion or redistribution (figure 3.4). Consequently, some species are expected to be very sensitive to climate change with consequent range reduction; others will be able to tolerate changing climate. This type of analysis can target which species are most likely to be sensitive.

Global climate change is not only a future concern. With increases in temperatures around the world of an average of 0.5° C in the past one hundred years (IPCC 1996), climate change has had contemporary effects on some wildlife populations. In the northern hemisphere, winter minimum temperatures have risen 3° C and spring maximum temperatures have risen 1.4° C since the middle of the twentieth century (Easterling 1997). More frequently, we are encountering signals that hint that climate change is already altering the distribution of wildlife. Studies that have compared range distributions over this time period shed light on the direction and extent of wildlife shifts in response to warming trends.

It appears that there has been a significant northern shift in geographic range for several species of birds, insects, and mammals. For instance, nearly two-thirds of the European butterfly species surveyed had shifted north (by 35–240 kilometers); only 3 percent shifted their range to the south. Likewise, birds in Britain and birds and mammals in North America have undergone significant northward range shifts in the past decades (Parmesan and Yohe 2003).

The Pace of Change

Although some species seem to be moving fairly rapidly, there is evidence that even for some highly mobile species such as birds and butterflies, the response to climate change is not as quick as would be expected were animals perfectly tracking shifts in climatic isotherms (Paremsan and Yohe 2003). For example, the European climatic isotherms have moved north 120 kilometers this century (+0.8° C), and a substantial portion of the nonmigratory European butterflies have shifted only a portion of this distance (Parmesan et al. 1999). A major constraint on migration may be availability of habitat. For example, habitat may not exist in contiguous patches due to urban development.

Whether or not human land development and/or natural barriers severely impact the ability of species to respond to changing climatic conditions will depend on the species' dispersal ability as well as the distribution of habitat and land use practices around the habitat. For example, it has been shown that for some species of butterflies dispersal will be seriously constrained because of habitat loss and habitat fragmentation, whereas other species of butterflies may not be so encumbered (Parmesan et al. 1999). We may also see evolution of morphologies designed to overcome barriers to dispersal (i.e., longer wings in some insect species; Thomas et al. 2001).

The above analyses address only one component of fitness: the ability to survive by seeking out more climatically favorable geographic locations. Without knowing the effects of climate change on reproduction, one cannot conclude that wildlife populations will remain viable simply because habitat is still available. Likewise, we cannot conclude that wildlife species will exhibit the same life-history traits under a doubling of atmospheric CO_2 as they do under current climate regimes. Indeed, mounting empirical evidence suggests that climate warming is altering the timing of life-history events. Examples comparing historical to current records has indicated shifts in breeding dates, body mass, and migration that are concurrent with increased mean spring and winter temperatures. Many common bird species have shown significantly earlier breeding dates ranging from a week to a month in advance of previous dates (Parmesan andYohe 2003).Amphibians in the United Kingdom now breed two to seven weeks in advance of breeding schedules from earlier in the century. Insects are passing through larval stages faster and are becoming adults earlier. This may have particular importance for the expansion of pest species ranges, and the extent of damage they are able to inflict on agricultural crops and on naturally occurring species.

Changes in the timing of life-history events due to climate change may not necessarily lead to species declines, however. We need to view these events in an environmental context to determine the important implications these life-history shifts may have for the populations and communities that are involved. For example, amphibians that are able to breed earlier may be relieved from a serious summer bottleneck (pond drying), allowing more individuals to metamorphose into adults. The effect of climate change may be that there is greater population recruitment than under current conditions. Alternatively, it is also important to realize that any benefit of earlier breeding may be offset if pond drying also occurs earlier in the season, or if biotic shifts such as abundance of important prey species cause environmental conditions to become less favorable. Also, a shift in earlier breeding dates for birds may allow a greater proportion of the population to produce multiple clutches during the breeding season, or to take advantage of new insect prey that are only available at earlier dates in the season. Thus uncommon species may become highly abundant.

The important point is that strong scientific evidence shows that human activity is causing mean temperatures to rise globally. Such change stands to impose strong natural selection on species. Some species may mitigate those environmental changes by migrating to more favorable climatic conditions but others will be constrained. Regardless, all species are being forced to adapt to the changes through selection on life-history traits and dispersal abilities. Such evolutionary response has been part of the natural cycle of life on earth for millennia. The difference now is the rate at which the environment is changing relative to change in our geological past. Such rapid change imposes very strong selection on species, many which may not have the genetic capacity or life history traits to respond quickly enough, if at all. Eventually natural selection may become so strong that individuals may fail to survive or reproduce altogether. This may precipitate one of the highest extinction events ever witnessed on earth (Thomas et al. 2004).

4

Ecological Limits and the Size of Populations

THE ECOLOGIST AND CONSERVATIONIST GRAEME CAUGHLEY (1976B) WROTE that all problems in wildlife management and conservation fall into one of three categories: (1) too many (overabundance), (2) too few (threatened and endangered), and (3) too many harvested. But, this begs the question: Too many or too few relative to what reference? In many cases, such as for example the problem of overabundance of North American white-tailed deer (*Odocoileus virginanus*; McShea et al. 1997; Cote et al. 2004), the frame of reference often involves the biological properties of the population intertwined with human perception (Sinclair 1997). As a consequence, scientific principles sometimes are muddled by differing human values, which, in turn, can cloud the policy debate about what management actions to take.

For example, to those who wish to avoid car accidents or the destruction of ornamental garden plants, even one deer may be too many. Those who oppose killing deer to reduce their population size, counterargue that the presence of one or a few individual deer within a local area is insufficient evidence to claim an overabundance problem. Either way, such values may or may not have any relation to the number of deer the natural environment can support. In order to have productive policy debates, we need to disentangle biological principles from value judgments. This requires first understanding basic principles of population dynamics and factors that may limit population size followed by application of these principles to reconcile human values of abundance with the biological capacity of the environment to support a population.

Simple Population Growth

Most species are capable of reaching huge abundances. Typically, however, they do not. Why? To answer this question, we need to consider the fundamental properties of population growth.

In discussing the life-as-a-game metaphor in chapter 2, I noted that the point of the game is to contribute as many descendants as possible to future generations. Thus, if there are two phenotypes that differ in their rate of survival and reproductive output (fitness), natural selection will favor the one that has the higher net survival and reproduction. This is because that phenotype produces genetic copies of itself at a higher rate than the other phenotype. Fitness, then, is a measure of that rate of increase in the abundance of a phenotype.

This kind of population growth is known as geometric or exponential growth because the number of individuals in the population multiplies rather than adds over time. This simple geometric process is what gives populations the potential to reach prodigious abundances.

Suppose we now counted the number of individuals in a population of a phenotype after each generation and plotted that number against time (figure 4.1a). This would produce a curve that begins slowly but then rises very rapidly. This kind of population growth is known as geometric or exponential growth because the number of individuals in the population multiplies rather than adds over time. This simple geometric process is what gives populations the potential to reach prodigious abundances.

Future population size can be forecast if we have two bits of information: (1) the population size at some starting time (call it time zero); and (2) the net rate of offspring production of an average individual in a population from one time period to the next, that is, a measure of mean population fitness. These two bits of information can be put into a mathematical equation describing a simple geometric growth process:

$$N(t) = N(o)e^{rt}$$

where $N(t)$ is population size (numbers) at some future time t, $N(o)$ is initial population size and r is the net rate of increase and e is the base of the natural logarithm (i.e., $e = 2.71828$). This equation may seem a bit daunting. But the reality is that it is used in everyday life: It is the equation used

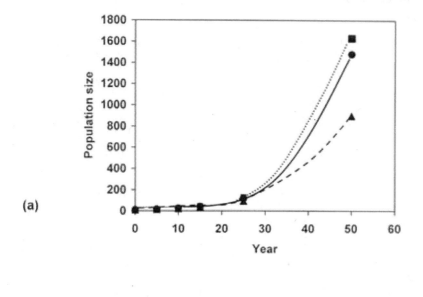

(a)

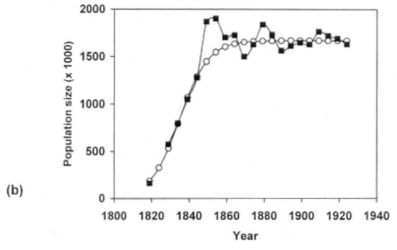

(b)

Figure 4.1. (a) Example of unbounded exponential population growth according to the equation $N(t) = N(0)e^{rt}$ where $N(t)$ is population size (numbers) at some future time t, $N(0)$ is initial population size and r is the net rate of increase. Solid circle represents the baseline where $N(0) = 10$ and $r = 0.10$, solid square demonstrates the effects of increasing $N(0)$ to 11 and solid triangle demonstrates the effect of decreasing r to 0.09. (b) Populations rarely continue to increase unbounded. The example of population growth of domestic sheep introduced to Tasmania reveals that populations can level off at some upper maximum size. The squares represent original census data points and the circles are calculated from a statistical fit through the census data. Original data are presented in Davidson (1938).

> As population size rises, that share diminishes in proportion to the number of individuals in the population. So, the higher the population size, the greater the intensity with which individuals must compete. Thus, rising population density continually feeds back to decrease individual fitness.

to calculate how invested money "grows" because of compound interest that is accrued. So, $N(t)$ could be the amount of money in an account after some fixed time t, based on the initial principle or deposit $N(0)$, and the annual percentage rate (APR) or interest rate r.

Let us now explore some properties of population growth using this equation. Suppose that we had a species with an initial population size of ten individuals, that the species breeds once each year, and that the mean fitness was such that the population grew at a rate of 10 percent per year. Population size after five, ten, fifteen, twenty-five, and fifty years is respectively 16.5, 27.2, 44.8, 121.8, and 1484 (figure 4.1a). We can change the conditions from this baseline by increasing initial population size by one individual. This translates into 148 more individuals than in the baseline population by year 50 (figure 4.1a). We can decrease the mean fitness in the population such that the population grows only by 9 percent. This results in 584 fewer individuals than in the baseline by year 50. The lesson here is that species populations can appear to be persisting at low densities for some time. Then, seemingly out of nowhere they can become highly abundant. Small changes in starting conditions can lead to dramatic difference over the long run. Over long-enough time periods, the geometric growth process can lead to prodigious numbers of individuals.

Species populations do not, however, continue to increase in abundance indefinitely. Eventually population size tends to level off at some upper bound (e.g., figure 4.1b) because individuals in the population must compete for a finite amount of resources or space. But how exactly is it that competition holds population abundances constant?

Individuals residing in a location must vie for their share of space or for their share of resources. As population size rises, that share diminishes in proportion to the number of individuals in the population. So, the higher the population size, the greater the intensity with which individuals must compete. Thus, rising population density continually feeds back to decrease individual fitness.

We can better understand the feedback process if we decompose fitness into its two fundamental components: per individual survival (or rather its opposite, mortality) and birth. When population size increases, per individual birthrates will decline because there are increasingly fewer resources available per individual to allocate to reproduction. Per individual mortality rates will rise with increasing population size, again due to declining abundance of resources that would normally be allocated to survival. (See figure 4.2a.) Whenever birth or mortality rates change with or depend on population size within a location, we say that the population is undergoing *density-dependent* growth. (Population size within a geographic location is called population *density*.) By contrast, fitness of individuals in populations undergoing the unbounded exponential growth described above is not influenced by population density. Such populations are said to undergo *density-independent* growth.

Ecological Balance and Carrying Capacity

In the face of intensifying competition with rising population size, per individual net fitness eventually will be reduced to the point where the per individual birthrate is exactly offset by per individual death rate (figure 4.2a). At this point, individuals are merely replacing themselves over their lifetime and the population will neither grow nor shrink in size. Instead it will remain at a steady state. This steady state is called *equilibrium*. The reader may be more acquainted with the vernacular term for equilibrium, "balance of nature." The problem with the term *balance of nature*, however, is that it gives the impression that there is a single natural balance. Yet, several different populations could exist at their own unique balance in different geographic locations. Hence, the idea that life on earth is in a single balance of nature is a popular but unfortunate misconception.

We have now explained how environments limit species populations. The equilibrium population size is effectively the fixed maximum population size that can be sustained or "carried" by the limiting supplies of resources or space in a geographic location. In ecology, this is formally known as *carrying capacity*, usually labeled K. In figure 4.1b, K for the Tasmanian population of sheep is on the order of 1,670,000 individuals. Carrying capacity is synonymous with equilibrium *for a single population*. Note the

> At this point, individuals are merely replacing themselves over their lifetime and the population will neither grow nor shrink in size. Instead it will remain at a steady state. This steady state is called *equilibrium*.

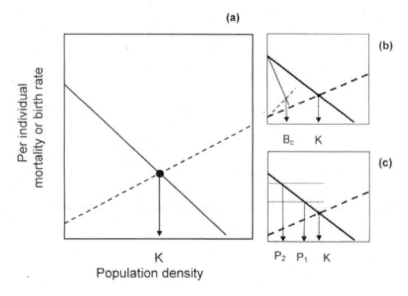

Figure 4.2. (a) Change in per individual birth or mortality rate as a function of increasing population density of a focal species. Birthrate (solid line) declines with population density because fewer resources are available on a per capita basis to allocate toward offspring production. Mortality rate increases with population density because fewer resources are available to support individual survival. The intersection of the two lines represents the point at which birth rate balances mortality rate—equilibrium or carrying capacity (K). This is where the population growth levels off after a period of increase (see figure 4.1b). (b) Competitor species can decrease a focal species' birth rate and increase death rate leading to a new equilibrium (B_c) in which the focal species is limited below its carrying capacity K. Predators can have the same qualitative effect as competitor species if they scare prey and thereby cause prey to spend less time feeding because they must be vigilant. This in turn reduces resource intake, which can lower birthrates and increase mortality rates across all prey densities. (c) Predators can increase prey mortality rate by capturing them (dotted lines) leading to a new equilibrium level that depends on whether or not the predator is inefficient (P_1) or highly efficient (P_2).

emphasis on single population here. The reason that we must be careful to link the term *carrying capacity* with single population dynamics will become clear when discussing the effects of predator and competitor species, and the policy implications of carrying capacity.

If it is true that populations must live within the confines imposed upon them by the environment, what happens when populations exceed their carrying capacity? Essentially, what happens if population density somehow ends up to the right of the equilibrium point in figure 4.2a? In such a case, mortality rates will exceed birth rates for all densities to the right of the equilibrium. Consequently, population density will decline until a balance between birth and mortality rate is recovered. The rate at

The equilibrium population size is effectively the fixed maximum population size that can be sustained or "carried" by the limiting supplies of resources or space in a geographic location. In ecology, this is formally known as *carrying capacity*.

which the population declines will depend on the difference between mortality rate and birth rate. Thus populations that far exceed their carrying capacity will crash faster than populations that only marginally exceed their carrying capacity.

But, if there are finite limits on population size, why can a population overshoot or exceed the equilibrium to begin with? Populations can overshoot their equilibrium because of lag effects. That is, at a certain population size, individuals will reproduce at a certain rate that is high relative to death rate. This is because individuals in that population do not yet encounter fierce competition for resources, so birth and mortality rates haven't been adjusted by strong competition yet. However, after a cycle of high births and comparatively low mortality it may be possible that the net number of individuals in the population have temporarily exceeded the level set by finite limitations. At this point, competition kicks in, increases mortality relative to birth, and causes population density to decline. As pointed out above, the rate at which population density declines is directly related to the degree of overshoot. For example, see the population data for Tasmanian sheep presented in figure 4.1b.

The same lag effect that caused a population to overshoot its carrying capacity can also lead to a population subsequently undershooting its carrying capacity. This then leads to a subsequent overshoot, followed by undershoot, and so on. The degree of subsequent overshoot will depend on the previous level of undershoot and the average fitness of individuals in a population, as determined by the way birthrate and mortality rate varies with population density. So, the overshoot and undershoot may eventually even out. Over time, the population would reach a steady state density set by its carrying capacity (figure 4.3a). The long-term dynamic exhibited by this level of lag effect is known as *damped oscillation* (Edelstein-Keshett 1988).

Populations can, however, exhibit other kinds of dynamics. For example, if per individual birth rates were initially slightly higher than in the previous case but declined more rapidly with population density, then the lag effect would be larger than in the previous case. Consequently, the pattern of overshoot and undershoot could be sustained indefinitely causing persistent oscillations as in the case when the population oscillates about the equilibrium reaching the same maximum density and alternately the same minimum density within each cycle period (figure 4.3b). This is known as a *stable limit cycle* (Edelstein-Keshett 1988). Technically, the population is in an equilibrium state even though it is oscillating—a stable limit cycle also represents a type of "balance of nature."

Stable limit cycles can take on many forms depending on the relative difference in magnitude between per individual birth and death rates, and the rate at which birth and mortality rate changes with population density (e.g., figures 4.3b and 4.3c). These cycles each are produced by identical intrinsic properties. They simply represent different variants of equilibrium. The implications of this variety of dynamics were used by Caughley (1976a) to propose some sobering "what if" questions about proposed management solutions to a wildlife overabundance issue.

Limit Cycles and the Management of the African "Elephant Problem"

In parts of Africa, there has been considerable attention drawn to the damaging effects of elephants (*Loxodonta Africana*) on forests. Elephants ring bark or fell mature trees and consume regeneration, which can lead to conversion of woodlands to open savannah or grassland—called the "elephant problem" (Caughley 1976a). The problem had been attributed to two causes. First, human-caused habitat loss compressed otherwise widely roaming elephant populations into local areas (the population compression hypothesis). Second, environmental change had allowed elephants to undergo rapid geometric increase (the population eruption hypothesis). One obvious management solution would have been to cull elephant population levels to the point where forests could regenerate.

Caughley critiqued these explanations on two grounds. First, both are effectively hypotheses that are predicated on the untested assumption that in the absence of human disturbances elephant populations and forests exists in a stable ratio of abundances over the long term. That is, they exist at a fixed-point (nonoscillatory) equilibrium. Second, the time trend under the two proposed hypotheses indicated that the population increase was re-

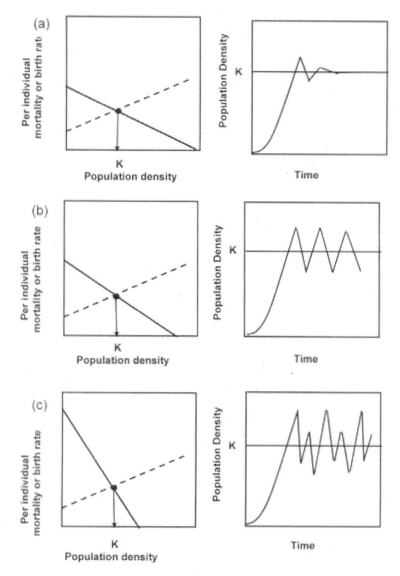

Figure 4.3. Effects of increasing the difference between density-dependent birth rate and mortality rate on the dynamics of a population. (a) Damped oscillation, (b) two-point stable limit cycle, (c) four-point stable limit cycle.

cent (within the last eighty years). Caughley instead proposed that elephants actually undergo stable limit cycle dynamics. Using tree ring counts of long-lived baobab trees, Caughley reconstructed the age-structure of the tree population. He noted that the age-structure was sharply bell-shaped and that the most abundant trees were 140 years old. This age-class of baobabs may have arisen when elephants were low in abundance. By corollary, when elephants were at the peak of their cycle, baobabs would be rare because they would have been highly exploited by elephants. Thus, if elephants and baobabs undergo limit cycle dynamics, the time for one complete cycle (peak to peak) would be, on average, 280 years. Allowing for some asynchrony between elephants and baobabs, a more conservative estimate of the cycle time is on the order of two hundred years.

Caughley was careful to note that this hypothesis, although plausible based on the data, would require further testing before ascribing a high degree of reliability to it. Nevertheless, the study illustrates that one can reason through alternative hypotheses using simple ecological principles. More importantly, if the limit cycle hypothesis is correct, then undertaking a cull (solution under alternative hypotheses) could seriously perturb the long-term dynamics of the system, perhaps leading to collapse of the managed population. Caughley's point is that one can and should take precautions to think through the alternatives before undertaking management actions that could have serious irreversible effects if they are based on misguided presumptions.

Competitors and Predators

The preceding discussion of population dynamics assumed that the only factors limiting the maximum size of a population at some location was competition for fixed supplies of resources or space. That is, I assumed that the factors influencing the dynamics of the population were *intrinsic* to the population itself. However, this assumption may not always be true. In reality, all species populations face *extrinsic* limiting factors because they are embedded in food webs in which they must compete with individuals of other species for resources or space. Moreover, they also may be subject to predation.

We can deduce the effect of competitors and predators on a focal species population using the fitness principles underlying figure 4.2a. In essence, individuals of a competitor species can have the same qualitative effect on individuals within a focal species population as individuals of the same species have on each other. That is, by vying for resources or space with members of another species, the per capita share of resources again declines. This decline in resource share causes an overall decline in per capita birth rates and increase in per capita mortality rates of the focal species (figure 4.2b). The consequence is that the focal species population may reach a new balance (B_c) between birth and mortality rates that depends on the density of the competitor species (the extrinsic factor). The population density at this new equilibrium will be less than that in the absence of the competitor species (figure 4.2b). That is, the competitor species (external factor) limits the focal species population below its carrying capacity.

Predators affect focal populations differently than competitor species. Predators can scare prey, which causes prey to spend less time feeding because they must be vigilant. This in turn reduces resource intake, which can lower birth rates and increase mortality rates across all prey densities. This would have the same qualitative net effect on per capita birth and mortality rates as competition and thus reduce the focal species below its carrying capacity (figure 4.2b). Predators also increase mortality of their prey by hunting and capturing them. The new equilibrium density of the focal species will fall below that in the absence of predators. The exact level at which the new balance is achieved will depend on whether predators are inefficient (P_1) or highly efficient (P_2) at capturing their prey. (See figure 4.2c.)

Weather

Point equlibria and limit cycles all result from fixed birth and death processes—called *deterministic processes*—that lead to order, even though there can be fluctuating or oscillatory behavior in the dynamics. All deterministically oscillatory populations cycle about an equilibrium but they do not reach that equilibrium exactly at any instant in time. Instead, they continually overshoot or undershoot the equilibrium. When they overshoot or undershoot, the density-dependent adjustments in fit-

All deterministically oscillatory populations cycle about an equilibrium but they do not reach that equilibrium exactly at any instant in time. Instead, they continually overshoot or undershoot the equilibrium. When they overshoot or undershoot, the density-dependent adjustments in fitness components (survival and reproduction) cause a directional reversal in population density. Thus the populations are continually drawn back or attracted to equilibrium.

ness components (survival and reproduction) cause a directional reversal in population density. Thus the populations are continually drawn back or attracted to equilibrium. Hence, we apply the special term *attractor* to equilibria of deterministically oscillatory systems (Edelstein-Keshett 1988).

Populations also can be influenced by *stochastic processes*. For example, suppose an unexpected storm passed through a location causing temperatures to plummet. Suppose some number of individuals died because they were poorly adapted to cope with such temperature change. The storm, an extrinsic factor, effectively increased the average mortality rate in the population.

Weather forecasters will tell you that specific weather events such as severe storms (or comparatively more benign events such as rainfall, dry conditions, etc.) are not highly predictable over the long term. Although we can count on regular seasonal changes in weather, we cannot expect that a storm event will hit the same location with the same ferocity and duration on the same date each year. There is considerable variability from one year to the next in rainfall, duration of winter, and so forth. Because the duration and intensity of this extrinsic factor is highly variable in space or time, it is unpredictable and so cannot be considered a deterministic factor. We call this a stochastic factor (Edelstein-Keshett 1988).

Stochastic factors also have important implications for population dynamics beyond simply altering mortality rates within populations. Suppose that individuals of a herbivore species rely on grassland forage to sustain themselves and reproduce. The nutritional quality of the grassland forage can be quite high in dry (good) years when the forage is cured by the sun's heat (except in the extreme case of drought when there is no forage and mass starvation ensues). In wetter (poor) years, the nutritional value is lower because the forage quality becomes diluted by water in the plant tissues. Because nutritional quality of the resource determines herbivore birth and death rates, it plays an important role in shaping the carrying capacity of the landscape. In other words, an extrinsic factor such as weather can limit the quality or abundance

> An extrinsic factor such as weather can limit the quality or abundance of food resources and thus determine the carrying capacity in any single year. Furthermore, the stochastic nature of weather causes the carrying capacity of the environment to fluctuate or oscillate up and down from one year to the next.

of food resources and thus determine the carrying capacity in any single year. Furthermore, the stochastic nature of weather causes the carrying capacity of the environment to fluctuate or oscillate up and down from one year to the next.

As a consequence populations will be forced to "track" these changes over time. Individuals experiencing "good" years will have high reproduction and low mortality leading to high net population growth rates. But, if the following year is a poor year, the population will have exceeded the carrying capacity set by environmental conditions the previous year (a lag effect) and so crash to a lower level. The fluctuation between good and bad years could also lead to oscillatory population dynamics. But, unlike a deterministic system, the dynamics of a stochastic system fluctuate in an irregular or random way. There is no single attractor for stochastic systems.

The potential for stochastic events to influence dynamics of populations creates a difficult empirical dilemma for managers. Populations whose dynamics are governed by deterministic processes are comparatively easy to manage. The nature and strength of cause-effect relationships are easily discerned and predicted. So managers can manipulate deterministic factors to achieve desired ends. Stochastic factors throw a proverbial monkey wrench into management by increasing uncertainty.

Carrying Capacity and Population Overabundance

We have all probably experienced, read about, or seen on television cases where a state wildlife management agency is called in to thin a deer herd because it is said to be overpopulating a local area. The act of thinning the herd by hunting evokes many different feelings in society. These feelings often boil over into highly newsworthy protest events or injunction hearings in court because defenders of animal rights often pit their interests against the interests of defenders of the hunting fraternity and those suburbanites who are tired of deer eating up their expensive ornamental plants. From a management standpoint, it is a political problem that needs to be dealt with quickly, usually by rapidly thinning the herd.

This action provides only a proximate solution to the environmental problem. It may result in a temporary fix that will require managers to thin the herd once the population again reaches high numbers. To solve the problem, we need to look deeper into the reasons for the problem in the first place.

Let's consider a hypothetical case in which a state environmental management agency is charged with managing a public forest preserve.

Management must balance the interests of a wide range of constituents. Some constituents use it for sports activities or recreation such as swimming. Some use it for observing wildflowers, songbirds, and deer. Suppose that the reserve was managed in such a way that it attracted deer to that location. The deer residing there would obviously consume some of the vegetation. Resource consumption, in turn, leads to births and the population begins to grow, putting increased pressure on the vegetation. The management agency becomes concerned because there is documented scientific evidence (e.g., McShea et al. 1997; Cote et al. 2004) that dense deer populations can prevent forest regeneration by eating tree seedlings. Deer can also alter the vegetation such that songbirds no longer reside in an area. Thus, important services of the nature preserve (e.g., forests, wildflowers, and songbirds) sought by the public stand to be jeopardized by a growing deer herd.

To head off this potential problem, managers look to other geographic locations that contain deer populations and try to come up with a representative number of deer that could be sustained locally in the preserve. Suppose the broad consensus was that the preserve should support no more than thirty deer. Based on this, management asserts that the carrying capacity of the preserve is thirty deer and then formulates all future management policy around that number.

However, it often happens that the deer population within a preserve is larger than this number or grows far beyond this number. Suppose that in our case that number was 120 individuals. By management's reckoning, we have a crisis because the deer population is exceeding its carrying capacity of thirty. Suppose that management went ahead and resolved this overabundance problem by culling the population back down to thirty animals and then left things alone. Often, we find that when this is done the population rebounds to its previous size of a hundred or more individuals within as short a time span as five years. This outcome then precipitates the next round of crisis management.

The problem is that this management crisis is probably more an artifact of policy based on superficial thinking than on the biology underlying population growth and carrying capacity. To understand what I am driving at, let's apply some of the principles laid out above.

The herd culling represents a perturbation intended to bring the population down to its predetermined carrying capacity for the preserve. In principle then, if the deer population was restored to its carrying capacity by management, we should see births in the population balance deaths (figure

4.2) and the population should remain at or near thirty individuals. The fact that the population can rebound to a hundred individuals within a short time span indicates that thirty is not the carrying capacity for the deer herd. One might counterargue that the population rebounds because of lag effects (figure 4.3). If this were true, then the population would eventually crash down in size on its own, because deaths would exceed births when populations exceeded their carrying capacity. Thus, culling would not be warranted—natural processes would take care of the overabundance problem.

There are a couple of reasons why an estimated carrying capacity of thirty is probably wrong. First, management obtained this number from other regions containing deer, not the particular preserve itself. As pointed out above, carrying capacity (equilibrium) can vary over space. So there is no guarantee that an estimate derived at one location applies exactly at another. Second, managing a public nature preserve is tricky business because it requires reconciling conflicting values of society. The choice of thirty for the carrying capacity in this case is more than likely based on management's *implicit* value to protect those many services of the preserve (e.g., songbird abundances or mature stands of forest) that are sought by society. This implicit value is in large measure the cause for the thinning of the herd.

Value Trade-offs and Ecological Implications

To disentangle science and values, consider the following set of concepts from Sinclair (1997). For the sake of argument, and to keep the concepts relatively straightforward, let's suppose that management must reconcile only two competing values: a large deer herd versus a densely wooded forest (implicitly carrying with it the attendant services of the forest such as songbird and wildflower species diversity). These competing interests are illustrated graphically where the axes represent the abundance of deer and forests in figure 4.4.

This graph immediately makes explicit the source of the management problem. Namely, that there is a trade-off in that management, and society for that matter, cannot have maximum values of all possible services within the preserve. At one extreme (the bold circle) we have maximum forest cover, but this can only be achieved if the deer herd is eliminated entirely. This is, effectively, a carrying capacity of the forest itself. If we allow deer to enter the preserve, then we must forego some forest cover because deer must utilize forest vegetation for sustenance. At the other extreme (the bold square) we have a maximum deer population size that the reserve can

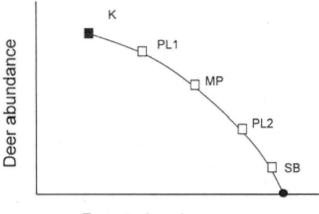

Figure 4.4. Decision curve representing different possible solutions to a management trade-off between maximizing forest abundance (and attendant ecosystem services), depicted by solid circle, and maximizing deer herd size (i.e., deer at carrying capacity K) at the expense of forest cover, depicted by the solid square. Intermediate solutions could range from modest deer population size with plenty of forest cover (SB) to modest forest cover and maximum offspring production rate (MP) of the deer herd. The figure also illustrates that predators do not determine carrying capacity of the deer: they limit deer population sizes below their carrying capacity. The degree of predator limitation (PL) depends upon whether predators are inefficient (PL$_1$) or highly efficient (PL$_2$) at capturing and subduing the deer. (After Sinclair 1997.)

sustain, that is there are sufficient resources in the forest such that deer births are balanced by deer deaths. This is the *biological* carrying capacity for the deer herd in the preserve. This, however, comes at the expense of a vast forest. Between these endpoints are many different combinations of deer and forests that are possible to achieve through management.

This line of reasoning now focuses the deer management issue on the appropriate explicit question: How many deer are we willing to allow in the preserve? Thus depicting the problem in terms of an explicit trade-off is a useful management concept because it now allows us to make our values explicit (Sinclair 1997). For example, suppose that society wants a high percentage of the preserve covered by forest so that it is able to view plenty of songbirds and such wildlife. This value might be achieved when deer population size is at or near a level depicted by the open square labeled "SB."

This is effectively the management scenario that I have been discussing all along. This is the target deer density that management needs to maintain to protect high levels of the other services of the forest preserve that society values. Thus, management must cull the herd on a regular basis to avoid the recurring boom-bust cycles arising from periodic culling. Management (an extrinsic factor) must lower the herd below its carrying capacity to achieve desired ends.

The utility of presenting information in this way is that it forces individuals with conflicting interests to make their values explicit. Management options can then be explored in ways that match ecological principles to value-oriented objectives.

This line of reasoning also allows us to address another very muddled concept that has been popularized by society. We hear that deer herds are now highly abundant because we have exterminated many of their natural predators such as wolves, coyotes, and bears. As a consequence, hunting is required as a substitute for missing predators in order to restore ecosystems back to their natural balance. The problem once again is that human values undergird the reasoning applied to the definition of "natural balance." According to the ecological principles laid out above, carrying capacity does not include mortality from predation because it is an extrinsic factor, not an intrinsic one. Predators limit prey populations below the prey population's carrying capacity. The extent to which this occurs depends upon the predator's hunting efficiency as depicted by the open squares PL1 and PL2. What this means is that loss of predators allows prey populations to increase from a former balance or equilibrium toward a new one: carrying capacity. What is implicit in society's wish to "restore the natural balance" is that society wants deer populations to stabilize at levels observed when predators were historically present, which is much below another balance—their biological carrying capacity. This suggests that we should abandon describing systems in terms of their "natural balance" and instead use the explicit terms we have learned such as carrying capacity and predator limitation.

This example illustrates how easily ecological science can become entangled with environmental advocacy. The reserve managers in my case example applied some ecological principles, but applied them only enough to advocate a certain set of values. As I pointed out in chapter 2, to remain objective and credible, ecological science must be presented in ways that honestly and clearly reveal the suite of management options to those interested in mediating environmental problems. Ecological science can do

---

What is implicit in society's wish to "restore the natural balance" is that society wants deer populations to stabilize at levels observed when predators were historically present, which is much below another balance—their biological carrying capacity. This suggests that we should abandon describing systems in terms of their "natural balance" and instead use the explicit terms we have learned such as carrying capacity and predator limitation.

---

this by presenting scientific insights in ways that reveal the trade-offs that different interest groups in the policy process must reconcile when making decisions and by illuminating the consequences of choosing one or another trade-off option. The trade-off curve clearly and explicitly depicts the alternative policy options and illuminates the consequence of choosing a particular option (i.e., the level of the deer population and the extent of forest cover in the reserve). Most importantly, the trade-off curve reveals that there is no ecologically "best" solution: there are just different states of an ecosystem that could be reached depending on what option is chosen. Thus, the favored solution depends entirely on different human values and preferences that must be reconciled through the policy process, not by ecological science.

5

Viability of
Threatened Species

THE FOLLOWING IS AN EXCERPT FROM AN ARTICLE PUBLISHED IN AN OUT-
door magazine. I am not providing the exact reference or identifying the
location where it happened because it would needlessly embarrass the
wildlife managers responsible for oversight of this resource. However, it is
a real-world example that requires some detailed scrutiny because this kind
of management activity is routinely carried out worldwide.

We Accept Your Apology. The Turtles, Alas, Do Not

When 222 baby hawksbill turtles poked their heads out of the sand at [such and
such] National Park last September, they were intent on doing what chelonian
hatchlings do best: bumbling down the beach and into the sea to join their
brethren, only a few thousand of whom survive (a statistic that ranks the hawksbill as one of the world's most endangered species). Sadly, a number of them
never got that far. Seven weeks earlier, a group of earnest volunteers had covered the fragile eggs with wire to shield them from predators—and then failed
to remove the protective cage before the hatch, which began a day earlier than
anticipated. By the following afternoon, 37 of the newborns had been toasted
to death by the sun. . . . "We all have our screw-ups," sighs the park's resource
management chief . . . "But this is the most lamentable one to date."

What were the managers thinking? They were thinking exactly what
most people would if they understood the plight of many sea turtle populations worldwide. We have probably all seen on television (or lucky enough
to witness in person) the annual ritual of adult female sea turtles coming
ashore on sandy beaches to dig nest holes and lay their eggs; only to have
marauding nest predators eat up most of those eggs. From a human value

standpoint, it stands to reason that we should protect the eggs of these wonderful, defenseless species from their predators. And so, wildlife management agencies and volunteers often hold vigil on beaches during the egg-laying season and place protective wire baskets over turtle egg nests.

At face value, such management derives logically from natural history observation and the retroductive reasoning (see chapter 2) that predators must be the cause of population decline for sea turtles. Here, however, is an example of management that may have acted on an untested hypothesis. We must ask the critical question: Why do this kind of management in the first place? To answer this question it will be insufficient to use the machinery of classic population ecology presented in chapter 4. This is because classic population ecology assumes that all individuals in a population—young and old—live in the same habitat all the time. But, we know that this assumption does not apply to sea turtles where adults who spend most of their time in the ocean live apart from eggs and young on the beach. To answer the management question we need to consider each life-cycle stage specifically—that is, we need to consider the *population age structure*—when attempting to understand the dynamics of the turtle population.

Sea turtles are long-lived creatures. Some can reach eighty years of age or more. They first begin to breed around age twenty; a female can produce many hundreds of eggs in a single breeding season. Suppose we followed the fate of individuals from the one hundred eggs of a single nest using survival estimates provided by Crouse et al. (1987). The data show that most eggs are lost to predation, fungal infections, accidents, and so on before they even hatch. Upon hatching, individuals must rush out to the ocean. Most make it, but some succumb to accidents or predation (Box. 5.1). Once in the ocean, individuals are highly likely to live; and to live to an old age. This pattern holds for many other species of reptile and amphibian species (Pianka 1988). Natural selection places a large toll on the egg and hatchling stages of the turtles.

For a population to be stable or sustainable over the long term, a parent only needs to be replaced by one surviving offspring. Of the thousands of eggs that are laid by a single female sea turtle over the course of forty to sixty years, only two turtle babies need to survive to replace their mother and father. This then begs the question: Why do sea turtles employ a game-of-life strategy of laying hundreds of eggs in a clutch and laying so many clutches? They do this because the need to breed on beaches and to spend

Box 5.1 Survival of Loggerhead Sea Turtles

Example of a longitudinal data set from a cohort of 100 loggerhead sea turtle (*Caretta caretta*) eggs whose fate is followed over six stages of their life. The survival numbers can be used to calculate stage-specific survival rate (px) and to graphically portray a survivorship curve. (Data from Crouse et al. 1987)

Age Class x	Number Surviving Sx	Number Dying Dx	Survival Rate px
Egg/hatchling	100	33	0.67
Small juvenile	67	26	0.62
Large juvenile	41	14	0.67
Subadult	27	8	0.70
Novice breeder	19	4	0.79
mature adult	15	3	0.80

the rest of their life in oceans prevents them from providing extended care for their offspring. Thus, turtles produce many more offspring than one would expect to survive and the newborn individuals must survive the mortality gauntlet on their own. It is simply a numbers game.

If the strategy of dumping eggs and leaving them to their random fate is an evolved strategy, then we must question whether or not protecting eggs is indeed strategically the correct measure for sea turtle conservation. Be-

Of the thousands of eggs that are laid by a single female sea turtle over the course of forty to sixty years, only two turtle babies need to survive to replace their mother and father.

fore we can answer this question, we must gain some insight into principles of life cycles and structured population dynamics.

Life Cycles and Population Dynamics

Individuals of most species develop through several life-cycle stages before they become adults. There is considerable variety in the modes of life-cycle development among different taxa. For example, most vertebrate species start life as newborns then pass through juvenile, young prereproductive adult, prime adult, and old adult stages. In these cases, the young are simply miniature copies of older members of a population. Usually, they reside in the same habitats as older individuals. Insects such as grasshoppers can likewise have simple development where young individuals reside in the same habitats as older individuals and they are also miniature copies of their older counterparts. In these cases, individuals may pass through five development stages (called *instars*) before reaching adulthood. Other insects are born and pass through similar instar stages in the aquatic realm (e.g., mosquitoes, dragonflies) but they switch to terrestrial habitats when they emerge as adults.

There is also complexity in life-cycle development. The most celebrated examples, perhaps, are the amphibians that are born as tadpoles in the water. Tailed and legless young tadpoles look very different than older tadpoles that have developed legs and lost their tails. Even these older tadpoles look different than adults who often leave the aquatic realm and spend a good part of their life on land. Perhaps the most striking examples of complex life cycles are in Lepidoptera (butterfly and moth) species. Here individuals develop from eggs, become larvae (caterpillars), enclose themselves in a casing (pupae), and remain dormant for a period of time after which they emerge as winged adults.

The life cycles of species can be represented schematically (figure 5.1). Consider the case of the loggerhead sea turtle with the six age classes listed in Box 5.1: eggs/hatchlings, small juveniles, large juveniles, subadults, novice breeders, and mature adults. Figure 5.1 illustrates with arrows the transition of individuals from one age class to another, or the aging process. It also illustrates which age classes contribute toward reproduction. The letters as-

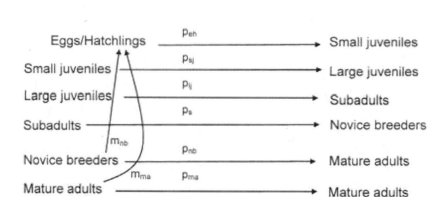

Figure 5.1. Depiction of life-cycle transitions and contributions of different stage classes to the overall population size and structure using loggerhead sea turtles as an example. The figure depicts life-history transitions of six stage classes (Eggs/hatchlings, Small juveniles, Large juveniles, Subadults, Novice breeders, and Mature adults). Horizontal arrows indicate that individuals in one stage class grow older during the course of a seasonal life cycle. Upward angled arrows identify those stage classes that produce offspring at the beginning of a season. The parameter m with stage-specific subscript represents stage specific reproduction; the parameter p with stage specific subscript represents stage-specific survival.

sociated with the arrows represent parameter values. The p-values represent the average probability that an individual will survive from the beginning of one age class to the next age class, called age-specific survival. The m-values represent the average contribution of an individual of a specific age class to the reproductive pool of the population, called age-specific fecundity.

Many analyses of population dynamics assume that here is one-year difference in age between the age classes in the population because many species undergo annual breeding cycles. Considerations of population dynamics based on such age structure are called *age-structured population dynamics* (Caswell 2001). Often, however, individuals from several different breeding seasons could be considered to be in the same life-cycle stage. For example, human teenagers comprise individuals from ages thirteen to nineteen; young adults are individuals say between the ages of twenty and thirty. In such cases, all individuals within an age class (e.g., teenagers) do not all become one age class older (e.g., young adults) over the course of a year.

Such is the case also for the loggerhead sea turtles. Consideration of population dynamics when different aged individuals can be grouped according to a common life-cycle stage is called *stage-structure population dynamics* (Caswell 2001). Both age-structured and stage-structured approaches are conceptually similar in the way they help to understand population dynamics. The difference between them is that a stage-structured approach accounts for the fact that some individuals remain within an age group for longer than one breeding period.

Ecologists and conservation biologists devote much effort to quantify age- or stage-specific survival and fecundity of species. This information is used to develop mathematical models that use this information to understand what the future population size and age or stage structure (number of individuals in each age or stage class) might look like in the face of various management regimes or environmental stressors.

Modeling Age-Structured Population Dynamics

Building a model of age-structured dynamics effectively is like building a spreadsheet with specific information about each age class within rows of a column. Each column then represents a subsequent year of life. Figure 5.2 shows how the different stage classes combine to influence population growth from time 1 to time 2. For example, only two stage classes (novice breeders and mature adults) contribute to the production of newborn offspring. They do so at the age-specific rate m. In natural populations it is usually the older, mature breeders that produce the lion's share of offspring because younger individuals, while physiologically capable of reproduction, may be too naive to breed or may have insufficient energy reserves to reproduce.

In addition to breeding, individuals of the different stage classes also become older. This is depicted in figure 5.2 by the downward-angled arrows. Individuals within a stage class grow older at rate p (with appropriate subscript designation for a stage class). If all of the individuals in a stage class survive to the next breeding season then p (the proportion surviving to the next age class) would be one or 100 percent. However, there are few if any populations (including humans) in which all members of one stage class survive to the next. Diseases and accidents all take their toll, independently of age—albeit younger or older individuals often are more likely to succumb to disease than are individuals in their prime. Thus, p for any age class is likely less than one. The lower the p value the higher the mortality risk for

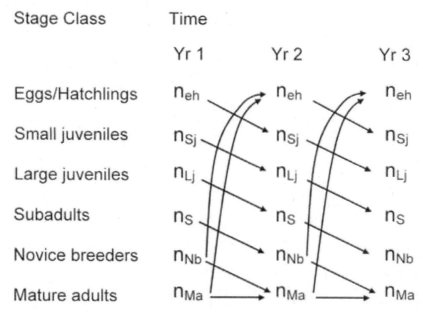

Figure 5.2. Hypothetical "spread sheet" for depicting aging and reproduction within six stage classes of loggerhead sea turtles over the course of three years. The value n with age specific subscript represents the number of individuals in the population belonging to a specific age class. The total population size can be calculated by summing all age-specific numbers in a column.

an age class. (Incidentally, these are the kinds of statistics that insurance actuaries use to calculate life or health insurance premiums.) This information can then be used to follow the fate of different-aged individuals over time (figure 5.2).

There is one problem with this modeling approach. That is, males in a population cannot produce offspring. So the model will distort dynamics if we include males in the current formalism. This issue is usually resolved by assuming that the sex ratio in a population is 50:50 (i.e., the number of males equals the number of females) and that females produce female offspring only. Under theses circumstances, we simply forecast the number of females in a population from one time period until the next and then double that number to forecast the total population size. The assumption of equal numbers of male and females is reasonable for most populations (Caswell 2001). When it isn't, we can build more complex models that ac-

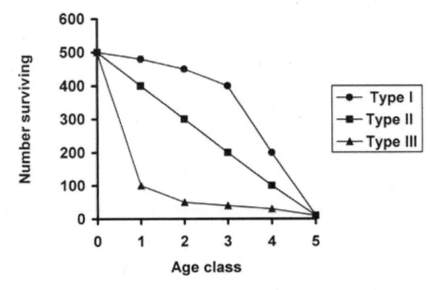

Figure 5.3. Examples of different survivorship curves exhibited by species. A Type I curve is representative of species that have high newborn and juvenile survival. These are typically species such as birds and mammals that engage in extended parental care. A Type II curve indicates that mortality rate is fairly constant across all age classes. A Type III curve applies to species with very high newborn and juvenile mortality rates and low adult mortality. These are typically species such as lizards and amphibians that produce many offspring in one breeding period and allow the offspring to survive the vagaries of the environment on their own.

count for sex ratio biases. The interested reader can consult Caswell for insight into building such complexity into structure population models.

In light of the above information, forecasting the numbers of individuals in each age or stage class from year to year requires that we know two things: (1) number of offspring produced by a female of a given age or life-cycle stage, (2) the probability that an individual will survive from one age or stage class to the next. There are two ways to get these data.

1. A longitudinal study in which a cohort of newborns is followed from their time of birth through their entire life. Age- or stage-specific survival and offspring reproduction is estimated from this cohort directly. This is impractical for long-lived organisms.

2. A cross-sectional study in which we collect data on the number of individuals of different ages in a population and their birth history. For humans, these data can usually be obtained from historic town records or from headstones on graves. With wildlife species, there are various techniques that have been developed to age individuals and obtain reproductive values (Schemnitz 1980). For example, hunter check stations run by State Wildlife agencies routinely are used to collect this kind of information from carcasses of hunted animals. These values are used to estimate age- or stage-specific survival and reproductive data. If we plot the values of population size against time (in this case, age is a surrogate for time) we can generate what is known as a *survivorship curve* (e.g., see Box 5.1).

For the data presented in Box 5.1, we find that the population size declines with a downward curve. That is, it declines rapidly at first and then slows its rate of decline with increasing age. This is not, however the only form of survivorship curve. Indeed, there are three general forms (Pianka 1988) known as Type I, II, and III survivorship curves (figure 5.3). Type I curves are characteristic of long-lived, large bodied species such as large mammals. Birds and some mammals display Type II curves. Reptiles, amphibians, and insects tend to display Type III curves (Pianka 1988). The example in Box 5.1 approximates a Type III survivorship curve.

Survivorship is environmentally determined. But, it is also a species characteristic that has been molded by evolution. For example, a Type III curve is often associated with species that cannot provide extended care for their offspring. Such species produce many more offspring than one would expect to survive and the newborn individuals must survive the mortality gauntlet on their own (e.g., sea turtles and predation). Alternatively, species that are able to provide extended care tend to produce fewer offspring and they invest more heavily in each of them to ensure that they survive the juvenile phase. This is important to keep in mind. Whenever we use management to change the survival probabilities of individuals in a species we are effectively imposing new selection pressures on these species

Whenever we use management to change the survival probabilities of individuals in a species we are effectively imposing new selection pressures on these species and hence are potentially playing with a strategy that has been molded by the species' evolutionary history.

and hence are potentially playing with a strategy that has been molded by the species' evolutionary history.

The survival probabilities (p) in conjunction with age class–specific birthrates (m) and numbers of individuals in an age or stage class can be used to calculate the survival and production of each age or stage class. For example, suppose two year olds survive to become three year olds with probability 0.8. Suppose an average two year old produces 1.2 female offspring (this means that if, say, there are 10 females and collectively they produced 12 female offspring, then on average there are 1.2 offspring per female). Suppose there are six hundred two year olds in the population that survive at a rate of 0.8. We can then calculate the number surviving to age three as 600 x 0.8 = 480. If the 600 produce 1.2 offspring each, on average, before they become one year older, then there will be 720 newborns that come from two year olds that age to become three year olds. These calculations can then be made for each age class. We can then calculate the number of individuals (n) in each age class (x) in a specific time period t ($n_{x,t}$) to understand population structure in a given time period. We can add up all of the $n_{x,t}$ values to calculate total population size in time period t (N_t). This can be repeated for each future time period based on data from the previous time period. Thus, if we wish to calculate population size fifty time steps into the future based on age structure, age-specific survival, and reproductive values, and population size in time 1, we simply iterate (i.e., repeat using data from the previous time step) through each calculation fifty times beginning with data from time 1. This task is routinely automated using computers that can calculate future population sizes in a matter of milliseconds.

Sensitivity of Populations to Disturbances

There are a number of things that we can do with this model. For example, suppose we were interested in the fate of a threatened species such as the Giant Panda (*Ailuropoda melanonleuca*). Pandas live predominantly in the Qinling Mountains in southwestern China (Zou and Pan 1997). They are specialist feeders on bamboo and their livelihood is threatened by habitat destruction (Zou and Pan 1975). To obtain key data, Zou and Pan (1997) studied the population for ten years. They discovered that females, on average, become sexually mature at age 3.5; males take two years longer on average maturing at age 5.5. Breeding occurs in March and April. Mothers give birth in August in a den. The reproductive interval, the time from one reproductive bout until the next, is about 2.5 years. This prolonged period between reproductive events is largely a consequence of mothers providing extended care for their offspring. Pandas can breed up to about sixteen years of age and they appear to display a Type II survivorship curve during the course of that lifetime.

In theory, species with extended birth intervals tend to have a slower capacity to recover from disturbances than species with less protracted birth intervals. In light of threats due to habitat destruction, the question facing managers is this: Can the Panda population sustain itself under current habitat conditions and demographic rates in the population? Zhou and Pan (1997) showed that the population could indeed continue to produce sufficient numbers of individuals to support positive population growth.

This estimate of the population's propensity to sustain itself can be viewed as a baseline reference. We can now enlist the machinery of structured population modeling to ask other questions that also aid in management planning. For instance, suppose that we wanted to know how sensitive the population might be to further habitat destruction. Let's suppose that habitat destruction meant that a female could no longer provision both herself and her offspring as well as under conditions of intact habitat. As a consequence, she and her offspring will suffer increased risk of age-specific mortality. The question then is: To which mortality rate is the population more sensitive, a mother's or the offspring's?

This question can be answered by systematically changing one parameter value at a time and then projecting how the population will grow after the parameter value has been changed. For example, suppose we estimated that a certain level of habitat destruction would lead to a decrease in cub survivorship from natural levels of 83 percent down to 70 percent. We can then reduce the parameter value p associated with cub survival from 0.83 down to 0.7, leave all other parameters the same as before, and then project what the future population would look like. In this case, we see that the population will persist, but over a two-hundred-year time span it will have two hundred fewer individuals than observed under higher survivorship (figure 5.4). If we change the survival of young, soon-to-breed females by a similar amount then we will see a much larger change in population size over time. Indeed, we will see a loss of 1180 individuals. This implies that the population is far more *sensitive* to changes in the survival of females that are reaching reproductive age than it is to cub survival. That is, conservation efforts should be devoted to protecting young females more than cubs.

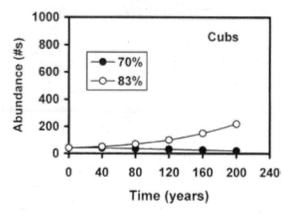

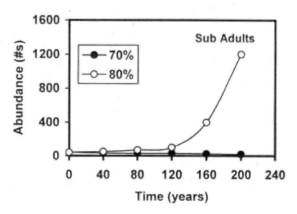

Figure 5.4. Sensitivity analysis of Panda population growth to changes in the survivorship of cubs or of prereproductive subadult females that stand to make a high future reproductive contribution to the population. Baseline data (open circles) indicate that the population can be sustained over the long term. Small decreases in survival (10 to 12 percent) should however cause the population to decline over time (solid circles). The analysis also reveals that protecting subadults will result in greater population abundances than protecting cubs. This suggests that conservation efforts should be targeted at maintaining high survivorship of older-age individuals. Graphs based on data from Zhou and Pan (1997).

When conducting sensitivity analyses of structured population models, population growth is often projected over a long (two hundred to one thousand year) time span. It is noteworthy that by doing this, the intention or the goal is not to provide an accurate forecast of long-term future population size. The intention, rather, is to provide a long-enough, stable time series so that one can obtain a reliable estimate of the degree of change in population size that may arise as a consequence of changing a parameter value by a fixed amount. The reason that one cannot assume the long-term projection to be accurate is that under natural conditions, parameter values do not remain fixed for indefinite time periods. As a consequence, structured modeling is normally used to conduct viability or risk analyses, rather than forecast population size (Beissinger and Westphal 1998).

Viability of Loggerhead Sea Turtles

We are now equipped with the concepts and tools to revisit our original question: Should management focus conservation effort at protecting sea turtle eggs on the beach? Crouse et al. (1987) did such an analysis for loggerhead sea turtles.

Box 5.1 summarizes the average stage-specific survival values of loggerhead sea turtles based on cross-sectional studies conducted in the field. In addition, nesting studies showed that novice breeders produce on average 127 eggs per breeding season and mature breeders average 80 eggs per breeding season (Crouse et al. 1987). These values are not known with exact certainty owing to wide variation about the average values. For example, egg survival can vary between 3 and 90 percent. Reproductive females can produce between one and seven nests per year. Given this variability, we can ask how sensitive the population is to a fixed percentage change in any one of the survival and reproductive values (Caswell 2001). Such a sensitivity analysis reveals where conservation should get its greatest "bang for its buck."

In the case of loggerhead sea turtles, the population is most sensitive to survival values rather than egg production. Moreover, the sensitivity is higher for older stage classes, which spend their life in the ocean. This conclusion derives from computer simulations that examine the effects of changing egg production and survival on absolute population growth rate. Like the case of the Panda, this analysis is accomplished by using average values to run a baseline computer simulation to estimate the long-term population trend. Baseline runs are then followed by simulation runs that systematically change one age-specific survival parameter value at a time.

In the case study, Crouse and colleagues systematically manipulated each survival probability individually by 50 percent and estimated population growth rate. Such an improvement is within the realm of possibility. This analysis reveals that for average baseline conditions observed in natural populations at the time of analysis, loggerhead sea turtles should have a negative population growth rate, implying that their numbers should dwindle to extinction if there is no management action taken to rescue the population. The analysis shows, however, that improving egg and hatchling survival by 50 percent *will not* reverse the decline. (See figure 5.5.) However, a 50 percent increase in the survival of small juveniles, large juveniles, and mature breeders will reverse the trend and lead to positive gains in population size. In essence, the analysis reveals that the pay off for concentrating efforts on improving egg and hatchling survival will not rescue the population, but improving survival of older-aged individuals may.

The analysis reveals that the pay off for concentrating efforts on improving egg and hatchling survival will not rescue the population, but improving survival of older-aged individuals may.

Rescuing Sea Turtle Populations

The simulation results beg the question: How does one improve the survivorship of individuals that spend most of their life unseen in the vast ocean? The key here is to identify the source of their mortality. It turns out that one important source of loggerhead sea turtle mortality in the ocean is being inadvertently caught in fishing trawler nets (Crouse et al. 1997). Unlike fish, which can extract oxygen from water through gills, sea turtles have lungs and so they must swim to the ocean surface regularly to breath. Turtles are unable to do this when they are caught in trawler fishing nets. The solution, therefore, is to figure out how to prevent turtles from being caught by fisheries. The ingenious technological solution is called a turtle excluder device, known by the acronym TED. This device selectively harvests fish and other seafood but prevents turtles from becoming entangled in the nets. However, TEDs also decrease the capture efficiency of nets and thus lower the income rate of the fishermen. Thus, there is a trade-off between an economic gain and conservation.

Further analysis illustrates how a structured modeling approach can evaluate different ways of reconciling this trade-off through different regulatory

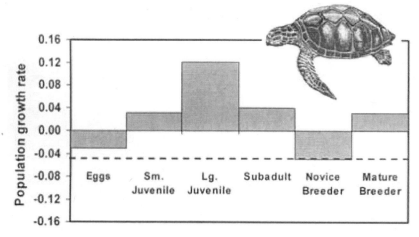

Figure 5.5. Analysis of the sensitivity of loggerhead sea turtle populations to 50 percent improvements in survival of different stage classes. The graph shows the effect of changing the survival of one particular age class (bars) on total population growth rate relative to a baseline (dashed line) that uses average parameter values (Box 5.1). For example, improving egg survival will increase the growth rate of the population from a baseline of −0.42 to −0.35. But, such improvement should not stem the population decline. Alternatively, improving large juvenile survival should lead to a positive population growth rate of 0.12. The analysis suggests that loggerhead population declines are best reversed by focusing conservation efforts at improving survival of large juveniles and subadults, not the egg stage. The graph is based on information provided by Crouse et al. (1987).

policy related to the use of TEDs (Crowder et al. 1994). There are two basic management options for the fishery. Fisherman might use the TEDs only during the shrimping season and only offshore. Or they could be forced to use TEDs year round and in all waters. An analysis of the options, using structured modeling revealed that loggerhead sea turtle populations would take seventy years or more to increase an order of magnitude in size if trawl fisheries used TEDs only during the shrimping season and only

The modeling does not provide evidence of cause-effect: The prescriptions are merely hypotheses formulated as management options. Thus, careful monitoring should follow any implementation of a management option to ensure that the modeling predictions are borne out.

offshore. If the trawl fishery were required to use TEDs all year round in all waters, then the population would reach the same population size in nearly half the time, assuming good compliance with regulations (Crowder et al. 1994).

This analysis serves as clear example how science might be interfaced with policy in efforts to make management decisions. There is one caveat. The modeling does not provide evidence of cause-effect: The prescriptions are merely hypotheses formulated as management options. Thus, careful monitoring should follow any implementation of a management option to ensure that the modeling predictions are borne out.

6

Biodiversity and Habitat Fragmentation

ECOLOGY AS A SCIENTIFIC DISCIPLINE WOULD BE HARD PRESSED TO CLAIM discovery of universal laws of nature, if in the strict sense, we require that a law explain the *invariant* causal link between a specific environmental condition and a specific ecological pattern or process (Lawton 1999). One ecological pattern that perhaps comes closest to being lawlike is the relationship between the size of an area and the diversity of species within that area, called the species-area relationship (Rosenzweig 1995). This relationship can be described qualitatively by a curve that rises and then saturates with increasing size of an area (figure 6.1a). But, the rate at which the curve rises and the level to which it saturates varies quantitatively among ecosyetms (e.g., arctic tundra vs. tropical rainforest), taxonomic identity of the species assemblage being examined (e.g., birds vs. insects), and habitat types (Rosenzweig 1995). Such quantitative variability would preclude the species-area relationship from being a true universal law (Lawton 1999). Nevertheless, the asymptotic trend is such a widely recurring empirical pattern (Rosenzweig 1995) that ecologists can at least be confident of ascribing some predictive reliability to this relationship (Peters 1991). Predictive reliability—in our particular case that increasing size of an area supports an increasing diversity of species, up to an upper limit—has profound implications for conservation and management (Peters 1986), which we will discuss below.

The prediction that species diversity saturates with increasing area applies only when we sample for what is called alpha (α) diversity (Rickleffs and Schluter 1993), meaning diversity within a specific area or habitat type, like an upland deciduous forest, a prairie grassland, the benthic zone of a freshwater pond, a salt marsh, and so on. The reason is that sampling across habitat types will cause the area affect on diversity to be confounded or washed-out by the effect of changing habitat and the different kinds of species associated with those habitats. This is not to downplay the impor-tance of diversity patterns caused by the mosaic of habitat types on a landscape, called beta (β) diversity. But, beta diversity requires scaling from local areas or habitats to the landscape, a subject that will be treated in detail in chapter 8.

One ecological pattern that perhaps comes closest to being law-like is the relationship between the size of an area and the diversity of species within that area, called the species-area relationship. Predictive reliability—in our particular case that increasing size of an area supports an increasing diversity of species, up to an upper limit—has profound implications for conservation and management.

Diversity Indices

There are various indices or measures of alpha diversity that are used by ecologists, each of which have different kinds of information content. The most commonly employed diversity index is called Species Richness. It is a simple count of the number of species in an area. This index gives equal weighting to all species, whether they occur frequently and thereby dominate an area or they are rare. Because it does not account for commonness or rarity, Species Richness can be conflated by the contribution of rare species to the measure of species diversity. In many cases, we want to understand the richness of species relative to their relative abundance.

Ecologists have proposed other diversity indices that combine Species Richness with various weightings for relative abundance. The first kind of indices, called *heterogeneity* or *diversity* indices (Krebs 1989), quantify either the likelihood that two individuals sampled randomly from an area are not the same species (Simpson's index), or the likelihood that one cannot predict to which species the next individual collected in an area belongs (Shannon-Weiner index). In both cases, larger values of the indices imply more heterogeneity, and hence diversity, than do smaller values of the indices. These two indices differ in their sensitivity to the weighting given to rare species. The Shannon-Weiner index is most sensitive to changes in the number of rare species sampled in an area whereas Simpson's index is most sensitive to changes in abundant species (Krebs 1989). Finally, ecologists have long known that natural communities have a few dominant species and many rare species and so wished to quantify such unequal representation.

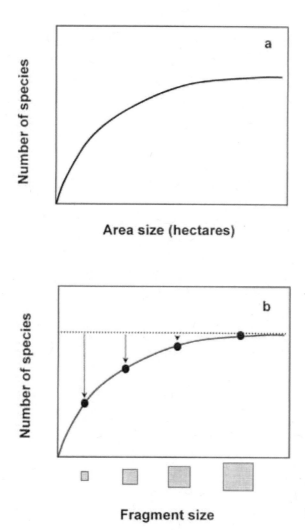

Figure 6.1. (a) The species-area curve describes the relationship between the number or diversity of species and the size of an area of habitat. Empirical evidence shows that the curve typically rises with area size and then saturates at an upper maximum. (b) The species-area relationship can be used to predict the effect of destroying large contiguous areas that leave scattered smaller habitat patches on the landscape. The length of the arrow indicates the magnitude of species loss caused by fragmentation within a fixed time frame. Larger fragments are expected to lose fewer species than are smaller fragments.

Evenness indices accomplish this by scaling the heterogeneity indices to a theoretical maximal value of diversity when all species are equally represented in the sample. In this case, large index values imply that the species are equally represented or equally abundant in a sample; small index values imply that there are a few species that are highly abundant and many that are rare. Ultimately, the choice of index to describe diversity depends largely on whether one is interested in emphasizing common species or rare species in an area.

The Ecology of Rarity

Each of these indices share one important feature: They are devoid of any biology other than measures and weightings of simple relative abundance. This can be problematic, especially for those indices that attempt to give greater weight to rare species. The problem arises because species rarity, in conservation, is often equated with being particularly fragile or threatened by humans and thus deserving of conservation concern. Yet there are many ecological reasons why particular species might be rare. So, viewing all rare species as a target for conservation concern can mislead conservation efforts (Rabinowitz et al. 1986).

Rabinowitz et al. identify seven different causes for species rarity based on contingency among three features of species populations: (1) their geographic range, (2) their habitat specificity, and (3) their local population size. Within this contingency, one might find that, at one extreme, a species has a broad geographic range but is rare within that range because it has very specific habitat requirements. A case example is the woodland caribou (*Rangifer tarandus caribou*), which occurs throughout the Canadian boreal forest but in extremely low population densities. This species is vulnerable to predation, competition with other ungulate species, and disturbances (Courtois et al. 2004). Woodland caribou especially prefer mature and over-mature lichen-covered conifer stands with irregular structure. These habitats are less suitable for other ungulates (Courtois et al. 2004). Moreover, such habitats tend to be widely dispersed on the landscape into isolated, small pockets that can only support low local population densities. Caribou also roam widely across the landscape as they move from pocket to pocket of habitat. This limitation on population size, together with the roaming behavior also reduces encounter frequency with natural predators (Courtois et al. 2004). Rarity due to a preference for a widely dispersed habitat type and migration among those habitat pockets may in this case simply be a consequence of a species adopting a strategy to deal with competition from other species and predation. At the other extreme, a species could have a small geographic range but have large local population sizes within that range. For example, a species of primrose (*Primula scotia*) occupies a tiny part of northeastern Scotland, but it has very large population sizes in several locations within that range (Rabinowitz et al. 1986). This pattern suggests that the species has substantial ability to thrive in a variety of conditions of soil moisture and fertility. Rarity due to this mechanism implies that there likely are some areas that are suitable to this species, but it has yet to colonize those.

The implication of this contingency is that efforts aimed at conserving rare species can be graded in their priority by using the three features of species populations as a decision tool (Rabinowitz et al. 1986). So, the most critical form of rarity—a species with small range, narrow habitat specificity, and small population size—is a species most deserving of immediate conservation attention because it is most likely to be jeopardized by disturbances that lead to loss of habitat (Rabinowitz et al. 1986). Species with other contingent combinations grading from narrow to wide geographic range, narrow to broad habitat specificity, and small to large population sizes become less of an immediate conservation concern.

Habitat Fragmentation and the Species–Area Relationship

Ever since Alexander von Humboldt's accounts of his expedition to the Amazon basin between 1799 and 1800 (Helferich 2004), we have revered the extraordinarily rich diversity of plant and animal life found in tropical ecosystems. It is no small wonder, then, that any human activity that imperils this diversity is likely to spark considerable attention.

In particular, vast expanses of humid tropical forests, which may harbor half of all species on the globe, are being lost or fragmented into very small areas as a consequence of logging and the use of fire for land conversion into agriculture (Ferraz et al. 2003). The species-area relationship predicts that habitat fragmentation qualitatively should lead to loss of valued tropical biodiversity, once fragments become small (figure 6.1b). So there is good reason for concern about the fate of tropical diversity in the face of this large-scale disturbance. But, the critical uncertainty for conservation is exactly what size and extent of habitat fragmentation is permissible without causing significant loss of species diversity.

Loss of Tropical Bird Diversity

The species-area relationship predicts that habitat fragmentation qualitatively should lead to loss of valued tropical biodiversity, once fragments become small.

To decrease this uncertainty, a team of ecologists (Ferraz et al. 2003) initiated in 1980 a large-scale fragmentation experiment in the heart of the Amazon basin near Manaus, Brazil, to evaluate the consequences of forest loss on biodiversity. The project created eleven isolated fragments that differed in size by orders of magnitude (two roughly one hundred hectare areas, four roughly ten hectare areas, and five roughly one hectare areas) with the matrix between the experimental patches comprised of cattle pasture (Ferraz et al. 2003). The study evaluated how quickly understory bird diversity, measured as species richness, disappeared from the fragments. The study employed standard bird mist net sampling for the ten-year duration of the experiment. The study discovered that smaller fragments tended initially to harbor fewer bird species than larger fragments, consistent with the species-area relationship (figure 6.1a). During the ten-year period, some bird species appeared to go extinct, some declined in abundance, and others remained stable. The rate of loss of a given proportion of species was higher in small fragments than in larger ones, again consistent with predictions of a species-area relationship (figure 6.1b). Ferraz et al.'s analysis of data also indicates that in order to increase by a factor of ten the time it takes for a fragment to lose 50 percent of its species, one must correspondingly increase the size of the fragment by a factor of one thousand. Even the hundred hectare (one square kilometer) fragments stand to lose half of their species within a decade or so. To guarantee long-term persistence of understory bird diversity in this tropical forest system for one hundred years, conservation efforts need to ensure that fragments do not become smaller than one to ten thousand hectares (ten to one hundred square kilometers).

Whether or not the insights from this research will be put into practice is beyond the scope of the science. The decision to require forest fragments in this system to remain larger than one thousand hectares to ensure long-term persistence of bird species diversity rests within the domain of government policy making. Government must also reconcile the interests of agriculture and logging with conservation. Nevertheless, the scientific study serves as an excellent example of how ecologists can, through careful experimental research at landscape scales, identify the cause and magnitude of risks to species diversity in ways that clearly illuminate the consequences of different policy options for habitat fragment size.

The Ferraz et al. study implicitly assumes that loss of species from a habitat fragment is a random event in the sense that each species of bird has an equal likelihood of going extinct and that it is only the loss of absolute living space that drives species in this system to local extinction. But, species do not live in "splendid isolation" of other species (Lawton 1999). They are embedded in food webs in which they are linked to competitor and predator species. In such cases, habitat fragmentation may disrupt important lines of dependency in those food webs and precipitate nonrandom extinctions due to altered competition or predation. A case example, which we now consider, is the attendant consequences of top predator loss from another ecosystem.

Fragmentation, Mesopredator Release, and Loss of Bird Species Diversity

Southern California has been transformed from a largely native sage-scrub chaparral landscape to a highly urbanized area. Vestiges of the native habitat are relegated to steep-sided canyons and small fragments within the urban matrix. As expected from the species–area relationship, bird species that normally depend on large, intact tracts of native sage-scrub habitat have undergone significant declines in population size or have become extinct (Crooks and Soulé 1999). Habitat fragmentation certainly is an important factor in bird species decline. But, closer inspection of the ecological processes underlying the dynamics of this system has revealed a more insidious cause for bird species decline.

In the original sage-scrub habitat, birds species were part of a food web (figure 6.2) in which the top predator, the coyote (*Canis latrans*), preyed upon several species of natural middle-of-the-food-web predators (mesopredators) including the striped skunk (*Mephitis mephitis*), the raccoon (*Procyon lotor*), and the gray fox (*Urocyon cinereoargenteus*), as well as more recent exotic predators (domestic cat, Opossum [*Didelphis virginanus*]). These mesopredator species in turn preyed upon birds' eggs, nestlings, and sometimes juvenile and adult birds. In this system, the coyote had an *indirect* beneficial

> Habitat fragmentation may disrupt important lines of dependency in those food webs and precipitate nonrandom extinctions due to altered competition or predation.

In this system, the coyote had an *indirect* beneficial effect on sage-scrub birds by *directly* controlling the abundance of the mesopredator species that prey upon the birds.

effect on sage-scrub birds by *directly* controlling the abundance of the mesopredator species that prey upon the birds.

The mesopredator release hypothesis (Crooks and Soulé 1999) predicts that effects of losing the top predator should cut off the indirect benefit the top predator provides to the birds (figure 6.2). This is because loss of the top predator releases control over the abundance of mesopredators. Unusually abundant mesopredators should then cause large declines in bird species populations and hence bird diversity. Coyotes require large expanses of habitat to thrive. Habitat fragmentation should precipitate the chain of events that lead to mesopredator release and devastation of bird diversity (figure 6.2).

Crooks and Soulé's survey of habitat fragments in the transformed landscape seems to bear this out. Fragment size was a very good predictor of mean coyote abundance. Furthermore, coyote and mesopredator abundance among fragments was inversely related. Finally, coyote abundance and bird species diversity (measured as species richness) was positively related among fragments.

Such surveys tend to be inductive in nature because they simply associate existing patterns of animal species abundance with existing patterns of fragmentation. However, Crooks and Soulé were able to ascribe potential causality by capitalizing on the fact that coyote visits to different fragments vary in space and time. They observed that changes in coyote abundance among habitats among the different years of the study led to changes in mesopredator abundance and bird species diversity in the ways predicted by the mesopredator release hypothesis.

Fragmentation thus stands to disrupt the nature of interactions among individuals within and between populations of species in a food web. In the example above, habitat fragmentation meant that coyotes were no longer able to reside within a single large tract of land. Instead, they were forced to wander among widely spaced pockets of habitat, thereby in-

Fragmentation thus stands to disrupt the nature of interactions among individuals within and between populations of species in a food web.

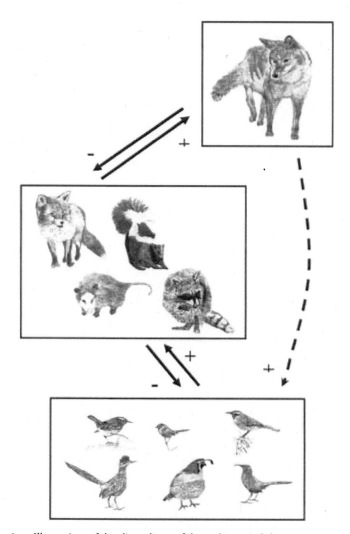

Figure 6.2. Illustration of the direct lines of dependency (solid arrows) among predator and prey species in the California sage-scrub food web examined by Crooks and Soulé (1999). The directions of the arrows indicate the direction of a species effect on another species. The minus sign indicates that a species has a detrimental effect on the abundance of another species. The plus sign indicates that a species has a positive effect on the population of another species. The dashed arrow indicates an indirect effect, an emergent property of food webs that arises when two remote species are connected by their common interaction with an intermediate species. Such emergent indirect effects are what create complexity in ecological systems by linking many species together in myriad ways. The implication of such interconnectedness is that disturbances do not necessarily affect a few species: Their effects can reverberate throughout the food web.

creasing their home range size. Ecological theory, which we consider next, predicts that such alteration in a species' ecology can affect the fundamental structure and long-term dynamics of species populations and ecological communities.

Habitat Fragmentation and Population and Community Processes

Populations of species that live in large tracts of intact habitat are expected qualitatively to undergo the kinds of population dynamics described in chapter 4. Namely, individuals of a population vie for their share of resources and the maximum limit on population size is determined by the capacity of the intact habitat to supply resources. Populations of species that live in patches of habitat that are separated on the landscape undergo different dynamics because they not only survive and reproduce within local habitat patches but they migrate among habitat patches on the landscape. This collection of local populations connected on the landscape by migration is known as a metapopulation (Levins 1969). A powerful metaphor for describing metapopulation dynamics is to liken habitat patches to lights on a Christmas tree in which patches occupied by the species are lit and patches that are unoccupied are dim. The ensuing dynamics can be imagined as the tree lights winking on and off. That is, some occupied patches become extinct due to random effects on small populations; however, some unoccupied patches are colonized by migrating individuals. In theory, the perpetual balance between patch extinction and recolonization allows the population as a whole to persist on the landscape even though it may not be present in any one patch at any one time (Kareivea and Wennergren 1995).

Fragmentation of large parcels of land into smaller, localized patches can thus change the fundamental character of population process because it forces individuals normally belonging to a large, contiguous population to be relegated to small, local populations scattered among small habitat pockets. An important feature of the population dynamics in fragmented habitats is that landscape-scale persistence and maximum size of the newly created metapopulation depends not only on the number and size of habitat patches but importantly on the ability of individuals to migrate among the local habitat patches (Kareivea and Wennergren 1995). Species will descend to extinction once the number of available habitat patches falls below a species-specific threshold. But, even if an ample number of patches is available, any development within the land matrix between habitat patches that hinders or impedes species migration could also be devastating to the population. The implication here is that well-intentioned habitat conservation aimed at protecting local habitat patches on a landscape may still paint a species into a proverbial corner of the landscape if a species' movement dynamics among habitat patches is not carefully considered.

An important feature of the population dynamics in fragmented habitats is that landscape-scale persistence and maximum size of the newly created metapopulation depends not only on the number and size of habitat patches but importantly on the ability of individuals to migrate among the local habitat patches.

Habitat fragmentation and consequent metapopulation process stand also to alter the fundamental structure of ecological communities. Within intact habitats, species that tend to thrive in high abundances are competitive dominants—those species that are best able to exploit local resources or preempt other species for gaining access to resources (Kareiva and Wennergren 1995). However, habitat fragmentation can change the playing field by strongly favoring mobility at the expense of competitive ability (Nee and May 1992). If we were to pit species with high competitive ability but low mobility (the competitive species) against species with low competitive ability but high mobility (the dispersing species) in a hypothetical game-of-life scenario, significant habitat fragmentation should shift the species dominance from the competitive type to the dispersing type. This is because the competitive species cannot colonize vacant patches faster than the rate at which it becomes locally extinct (Nee and May 1992). This is wholly counterintuitive because one would normally guess that highly successful, abundant species are the ones least at risk to human disturbance (Kareiva and Wennergren 1995).

Habitat Fragmentation and Extinction Debt

The effects of habitat fragmentation on the species composition of communities may not be immediate when species can be ranked in a competitive hierarchy from strong competitor but weak disperser to weak competitor but strong disperser. Under these conditions, lags can exist between the time a habitat is fragmented and the time when a species disappears altogether from all habitat fragments. In tropical forests, this lag can be between ten and fifty years (Ferraz et al. 2003); in California sage-scrub it can be up to seventy-five years (Crooks and Soulé 1999). Theory shows that the level of extinction rises exponentially with the degree of habitat loss. Initially small amounts of habitat loss should precipitate only low numbers of species extinctions. Greater amounts of loss can cause a sharp rise in extinction rate. As explained above, it may be the highly competitive species that are most susceptible to extinctions. However, highly competitive species tend to be highly abundant at the onset of habitat loss so it will be some time before their populations disappear. Species should also disappear sequentially in rank-order of their competitive ability. These factors together imply that fragmentation can precipitate a chain of species extinctions that are irreversible once the fragmentation process is initiated. Without some careful thinking, current habitat fragmentation may pass on a considerable legacy of debt—the "extinction debt" (Tilman et al. 1994)—the burden of which will only truly be realized by our grandchildren and generations beyond.

> Theory shows that the level of extinction rises exponentially with the degree of habitat loss. Initially small amounts of habitat loss should precipitate only low numbers of species extinctions. Greater amounts of loss can cause a sharp rise in extinction rate. Without some careful thinking, current habitat fragmentation may pass on a considerable legacy of debt—the "extinction debt."

We must be careful not to ascribe absolute truth to the assertions made by the various theories described above. After all, the predictions are only as accurate as the assumptions on which they are based conform to nature. We must ask several questions: Does a trade-off exist between competitive and dispersal ability among species? Is the likelihood of colonizing any patch on a landscape equal across all patches? Can these spatial processes be examined without considering the effects of predators on the competitors? Theoretical examinations that relax these assumptions suggest that the extinction debt idea is robust (insensitive) to changes in assumptions (Kareiva and Wennergren 1995). The ultimate arbiter, however, will be data derived from experimentally testing the predictions on large landscapes. Such experimentation may, however, be logistically challenging and even impossible on ethical grounds. Thus data to address this issue may not be easily forthcoming.

Nevertheless, the modeling presented here encourages some important pause for sober reflection before taking any action. This is because it forces one to ask critical "what if" questions. The insights from theory coupled with empirical evidence presented in this chapter begin to reveal that the consequences of human impacts can cascade through myriad pathways in highly interconnected networks of species. The consequences of these effects may take decades to centuries to fully play themselves out. Possibly it may take equally long periods, if ever, to reverse these effects.

7

The Web of Life: Connections in Space and Time

ALDO LEOPOLD, THE FATHER OF MODERN CONSERVATION ETHICS, DID NOT begin his career writing and speaking on the importance of conserving ecosystems. Ironically, his first profession after graduating college was as a predator control officer with the U.S. government. This job required that he eradicate predators from wilderness areas with the express purpose of enhancing the abundance of game species. This experience, however, left an indelible and formative impression on him as he observed during his lifetime the consequences of systematic predator removal. He articulates that impression in his essay "Thinking Like a Mountain" (1953), an essay that represents the beginning of ecosystem conservation ethics.

He opens the essay by recounting his hearing a wolf's "deep chesty bawl" echo throughout the mountain canyons. He continues with a reflection:

In those days, we had never heard of passing up a chance to kill a wolf. . . . We reached the [mortally wounded] old wolf in time to watch a fierce green fire dying in her eyes. I realized then, and have known ever since, that there was something new to me in those eyes—something known only to her and to the mountain. I was young then, and full of trigger-itch; I thought that because fewer wolves meant more deer, that no wolves would mean hunters' paradise. But after seeing the green fire die, I sensed that neither the wolf nor the mountain agreed with such a view.

Since then, I have watched state after state extirpate its wolves. I have watched the face of many a newly wolfless mountain and seen the south-facing slopes wrinkle with a maze of new deer trails. I have seen every edible bush and seedling browsed, first to anemic desuetude, and then to death. I have seen every edible tree defoliated to the height of a saddle-horn. Such a mountain looks as if someone had given God a new pruning shears, and forbidden Him all other exercise. In the end, the hoped-for deer herd, dead of its own too-much, bleach with the bones of the dead sage, or molder under the high-lined junipers. Just as a deer herd lives in mortal fear of its wolves, so does a mountain live in mortal fear of its deer.

Leopold poetically describes here the trophic cascade concept that I presented in chapter 2 in which the top predator in a food web has an indirect beneficial effect on plants by controlling the abundance of its prey species (figure 2.2a). This early-twentieth-century large-scale experiment in game management did not ultimately enhance game species abundance but rather unwittingly caused the collapse of an entire system. The essay also underscores another important consequence, namely the legacy that such management action passes on to future generations. Leopold notes that a deer lost to predation can be replenished after two to three years, whereas a range destroyed by highly abundant deer may take two to three decades, if ever, to be replenished.

> Predator control, an early-twentieth-century large-scale experiment in game management, did not ultimately enhance game species abundance but rather unwittingly caused the collapse of an entire system.

Ecosystems in Time

The legacy of predator eradication has been recounted more recently for a food web in riverine areas that comprise part of the prairie ecosystem in the western United States (Berger et al. 2001). Such areas were once dominated by large carnivores such as wolves (*Canis lupus*) and grizzly bears (*Ursus arctos*), large herbivores such as moose (*Alces alces*) and stream- or river-side (riparian) tree and shrub vegetation. In this case, individuals of each of these large mammal and plant species live many years (more that ten years) and so there is a considerable lag between the time that a predator species is lost and the attendant adjustments in the lines of dependency made by the remaining herbivores and plants.

The large carnivores that normally were part of this food web were lost between 75 to 150 years ago because of government policies to eradicate predators. This loss triggered a cascade of events with some indirect surprises that are only being realized in the last twenty or so years. It turns out that moose can have important localized effects on riparian vegetation by eating woody species such as aspen, willows, and cottonwoods (Berger et al. 2001). Indeed, they can be particularly damaging to the vegetation because they can reach high densities in areas where predators are absent. The gradual build-up of moose populations and corresponding decline in riparian vegetation has now resulted in the decline of numerous migratory songbird species, including species that depend wholly on riparian habitat for their existence (Berger et al. 2001). The sobering point here is that the early-twentieth-century government policy of carnivore extirpation has created a legacy of ecosystemwide effects that are only fully realized by the grandchildren and great-grandchildren of those originally engaged in the predator control efforts.

> The sobering point is that the early-twentieth-century government policy of carnivore extirpation has created a legacy of ecosystemwide effects that are only fully realized by the grandchildren and great-grandchildren of those originally engaged in the predator control efforts.

These examples illustrate that species diversity is represented not by members of a single trophic level but by food webs in which species in different trophic levels of the system are directly and indirectly connected to each other. This diversity is maintained because carnivores control the abundance of herbivores. This in turn prevents herbivores from overeating plants, allowing the entire system to remain intact. Removing the predators leads to loss of species diversity. Such effects can, however, be reversed through careful management aimed at restoring direct and indirect effects.

Restoring Interconnections in Ecosystems

Prior to the late 1990s, wolves had been absent from Yellowstone National Park for the better part of seventy years (Ripple and Beschta 2003). During that time, there was a herbivore-caused decline in the riparian vegetation, especially cottonwoods (*Populus sp.*) and associated woody plants, along the Soda Butte Creek and the Lamar River in the northeastern part of the park, similar in nature to that chronicled in the Berger et al. (2001) study described above. In this case, however, the dominant herbivore is the North American elk (*Cervus elaphus*) rather than the moose. Wolves were reintroduced to the park in the winter of 1995–1996 (Ripple and Beschta 2003) and within seven years, there were pronounced differences in elk browsing intensity and the height of riparian woody plants between sites with low visibility and comparative absence of escape barriers and nearby sites that were more open.

This difference was largely because elk avoided the low visibility areas putatively due to heightened predation risk in those areas. Moreover, the difference arose because young cottonwoods on high-risk sites were growing taller each year of the last four years, while they were not growing there prior to the wolf reintroduction. By comparison, there was little height change in the low-risk sites due to persistent browsing by elk. This case study demonstrates that a predictive understanding of trophic cascades (i.e., understanding direct and indirect connections) can lead to management action that hastens the recovery of an ecosystem, despite long-term absence of a key component of the ecosystem.

Enhancing Species in Ecosystems

Commercial and sport fisheries are an important economic activity worldwide. It stands to reason then that managers develop management programs that sustain or even enhance the production of economically important fish species. Nevertheless, attempts to enhance the production of a single species without considering the way the focal species is linked to other species in a system can lead to outcomes that are opposite to that desired.

Spencer et al. (1991) chronicle the saga of management aimed at enhancing the production of a landlocked salmon species, Kokanee salmon (*Onchorynchus nerka*), in several lakes that are part of the Flathead catchment in northwestern Montana. Managers introduced freshwater opossum shrimp (*Mysis relicta*) to several lakes between 1968 and 1975 believing that the shrimp would serve as a key supplement to the salmon's diet. The problem with this strategy was two-fold. First, it turned out the salmon did not eat the opossum shrimp. Second, opossum shrimp are voracious predators of zooplankton species that are a major food source for the salmon. Management effectively inserted into this ecosystem a species that turned out to be a strong exploitative competitor (figure 2.1c) of salmon. Moreover, because of their voracity, opossum shrimp flourished and reached peak numbers by 1981 (Spencer et al. 1991). This in turn caused the plankton species on which the salmon depended to be decimated, precipitating the collapse of the Kokanee salmon population and its associated fishery. By the early 1990s, the species composition of the lake ecosystems also became transformed. The top predators in the lake are now whitefish (*Coregonus clupeaformis*) and small lake trout (*Salvelinus namaycush*). The effects did not simply stop within the boundaries of the water bodies themselves. Many species of birds and mammals such as bald eagles (*Haliaeetus leucocephalus*), gulls (*Larus sp.*), grizzly bears (*Ursus arctos*), coyotes (*Canis latrans*), mink (*Mustela vison*), and otter (*Lutra canadensis*) depended on spawning salmon as a key food source. These species also declined in abundance in many of the tributary watersheds feeding the Flathead lake system in which the salmon spawn.

Ecosystems in Space: Linkages across Geographic Boundaries

The above treatment of ecosystem complexity shows the temporal legacy of human impacts on ecosystems. It also alludes to the fact that ecosystems cannot be viewed as though they were self-contained entities. For example, lakes—often viewed as being isolated from the surrounding terrestrial systems by a hard, shoreline boundary—are affected by seasonal runoff as melting snow in spring flows down hill slopes carrying with it nutrients into lakes. Such run-off in turn can be important in sustaining the structure and functioning of lake ecosystems (Pace et al. 2003).

The older conceptualization of an ecosystem being a self-contained entity is now giving way to recognition that ecosystems are connected to each other by flows across landscapes in which they are juxtaposed. This new way of thinking has been championed by the late Gary Polis who essentially asked the question: What if we focus on the consequences to food chain dynamics of the flow of externalities to a system rather than concentrate solely on the components within the system? Polis began asking this question after studying oceanic island ecosystems off of Baja California that are sharply separated from each other and the mainland by large distances and a seemingly impermeable salt water barrier. These arid islands provide a largely inhospitable environment. They are covered with *Opuntia* cactus, myriad species of flying insects, and their web-building spider predators (Polis and Hurd 1995). Curiously, however, the islands supported extraordinarily high densities of spider predators and this trend was more pronounced on smaller islands than on larger ones. This oddity ran counter to current ecological concepts. It is larger island and mainland ecosystems that are supposed to be better able to support absolutely more top predators. This is because they have higher in situ plant production to sustain those higher levels of the food chain. Polis and Hurd's observations instead suggested that the small islands held the more productive systems.

In searching for an explanation, the researchers noticed that the shoreline was not an impermeable boundary. Rather, there was a considerable abundance of nutrient rich resources in the form of algae and drowned animal carcasses that washed up onto the shore from oceanic drift. This resource input was sufficient to sustain insect species that consumed the algae and scavenged the decomposing carcasses, species that might not be as highly abundant if they had to rely solely on plant production on the islands themselves. In effect, the island economies received a subsidy from the ocean that eventually supported very high abundances of top predators. Moreover, the smaller islands were more productive because of the physical properties of their boundaries. Smaller islands have a higher perimeter to area ratio. That is the smaller islands have more shoreline relative to their overall area than do larger islands. This property allows consumers from all over the small island to access the subsidy. By contrast, individuals living in the middle of the larger islands have a lower likelihood of encountering the subsidy. The subsidies also influenced the dynamics of the island ecosystem. The abnormally high abundance of spiders led to an unusually high capacity to control the abundance on the island's herbivorous insects, thereby lessening the insect damage to plants. Thus, the effects of the subsidy reverberate through the whole island system. Shut the subsidy off and the ecosystem could collapse to a comparatively barren desert.

> The effects of the shoreline subsidy reverberate through the whole island system. Shut the subsidy off and the ecosystem could collapse to a comparatively barren desert.

The lesson from this study is that the two very different kinds of ecosystems can be inextricably linked through resource flows across their boundaries. The amount of subsidy provided and its attendant effects depend very closely on the dynamics of species interactions within each ecosystem. That is, if marine production is altered by environmental impacts or from species imbalances in the marine food chain, then the amount of subsidy to the island can become altered causing a cascade of effects on the island. It is the broader landscape and the ebb and flow of resources across ecosystem boundaries on that landscape that drives dynamics.

Species Link Ecosystems on a Continental Scale

The effects of subsidies can have more far-reaching effects by linking ecosystems in vastly different parts of a continent. Agricultural production in the wintering grounds of migratory waterfowl in the southern United States and along northward migratory routes, improves the health and survival of geese that spend the summer in Canadian arctic tidal flats. These geese have reached enormous densities such that they now have a cascading and devastating effect both on the production and abundance of vegetation and on nutrient cycling within that arctic ecosystem (Jefferies et al. 2004). In particular, they are beginning to cause the collapse of normal arctic ecosystem function and there are hints that the damage may be irreversible because of wholesale changes in the physical attributes of the soil (moisture, salinity, and temperature).

The most important lesson here is that it is no longer tenable to think that the world is divided by sharp, impermeable boundaries, be it between land and water, urban and rural, south and north, tame and wild. More importantly, we need to contemplate the effects of our local actions in the broader context. Returning to the goose example, reversing the trend of ecosystem damage in the arctic is not simple. Developing policies to eliminate the agricultural practices in the southern United States to protect comparatively uninhabited arctic ecosystems in another country would in all likelihood not be given more than cursory consideration. But, what if the opposite happened? What if some sort of management activity in the arctic enhanced the goose populations such that geese heavily damaged agricultural crops during their stay on the wintering grounds? This reasoning underscores that thinking globally when acting locally can no longer rest within the domain of environmental activism. It must now be a matter of course. This is further underscored by recent understanding of effects on ecosystems across global scales.

What if the opposite happened? What if some sort of management activity in the arctic enhanced the goose populations such that geese heavily damaged agricultural crops during their stay on the wintering grounds? This reasoning underscores that thinking globally when acting locally can no longer rest within the domain of environmental activism. It must now be a matter of course. This is further underscored by recent understanding of effects on ecosystems across global scales.

Humans and Geographic Transport of Species

Globalization through expanding transport and commerce over the past five hundred years has also increased the capacity of species within one geographic range to invade new geographic ranges (Mack et al. 2000). Such biotic invasions occur when organisms are transported to new, often distant ranges where their descendants proliferate, spread, and persist. Such invasions are cause for significant concern in the conservation of ecosystems and endemic biodiversity within them. Invading species can alter fundamental ecological properties such as the dominant species in an ecosystem, and the biophysical features of ecosystems, including nutrient cycling and plant productivity (Mack et al. 2000).

In the game of life, invasive species are, on average, better at utilizing the same set of resources as those used by a native species or they are better at translating those resources into fitness because they have fewer costs in their new geographic locations owing to release from biotic constraints. For example, predators might not recognize a novel invader as a potential prey item, thereby freeing the invasive species from the need to devote time or energy to avoid predators. In this case, the invasive species is less interconnected to other species than are the native species with which it competes. In other cases, the invasive species may be a novel, voracious predator that devastates native species that have not evolved the capacity to deal with this new predator species.

Invasive species often affect ecosystems by wedging themselves into them and then systematically causing the collapse of the entire system. A classic example of such an effect involves the Zebra mussel (*Dreissena polymorpha*). This species is native to the Caspian Sea region of Asia and was transported to the U.S. Great Lakes in ballast water in a cargo ship (Ricciardi and MacIsaac 2000). The ballast water was discharged before the ship entered port in Detroit. Adult female zebra mussels, which are no bigger than your thumbnail, can produce between thirty thousand and one hundred thousand eggs per year. Zebra mussels invade aquatic food webs by attaching to fixed surfaces. They then outcompete native species by rapidly filtering phytoplankton from the water column. The loss of the phytoplankton base has reverberated upwards in the ecosystems through loss of zooplankton and fish species in higher trophic levels. In addition, Zebra mussels

> In the game of life, invasive species are, on average, better at utilizing the same set of resources as those used by a native species or they are better at translating those resources into fitness because they have fewer costs in their new geographic locations owing to release from biotic constraints.

are able to sequester toxins and pollutants in high concentrations and then excrete them in feces leading to local concentrations of toxic chemicals, which further kills native life. This creates clear water zones in lakes that foster invasions by other species (Ricciardi and MacIsaac 2000). Zebra mussels are effective because they are potent exploitative competitors (drawing down phytoplankton) and interference competitors (production of toxics), two abilities shared by many other invasive species (Mack et al. 2000).

Interconnected Species as Conduits of Climate Effects

Recent evidence has shown that cyclic weather phenomena also can have long-term (decade- to multiple decade–long) impacts on species interactions in and properties of ecosystems. Analysis of forty years of annual data on a well-studied system of wolves, moose, and balsam fir (*Abies balsamea*) on Isle Royale, Michigan (Post et al. 1999) reveals that a cyclic weather phenomenon with a decadal trend in temperature, moisture, and winter snowfall—the North Atlantic Oscillation (NAO)—has a strong influence on ecosystem function.

The NAO phenomenon is determined by atmospheric pressure differences off of continental North America between Iceland and the Azores in the North Atlantic. But, the NAO has profound effects on the continent. Whenever NAO conditions are such that they cause high snowfall levels in northeastern North America, Isle Royale wolves become extremely efficient predators of moose that are encumbered by deep snow. Consequently, wolves reduce moose populations to levels where they cause very limited damage to balsam fir. Wolves thus play an important direct and indirect role in determining the compositional make-up of forest vegetation. This role of wolves becomes diminished when NAO causes snowfall levels to be low. In this case, moose easily evade their predators causing moose populations to abound and inflict considerable damage to balsam fir (Post et al. 1999).

The connection between climate and species interactions also has some surprising implications in the face of another long-term environmental change discussed in chapter 3: global warming. Certain environmental conditions realized under NAO forcing, namely winters with anomalously warm temperatures and little snowfall, are akin to the kind of conditions expected from global warming due to rising levels of the greenhouse gas CO_2. Over the long term, climate warming can cause a cascade of effects including declining wolf populations, rising moose populations, and declining balsam fir productivity. Moreover, moose may increasingly suppress sapling tree recruitment resulting in a more open forest canopy with a changed understory of shrub and herb species diversity. Thus a chronic human-altered environmental factor has the potential to transform the species make-up of large ecosystems by decoupling important direct and indirect linkages among species (Schmitz et al. 2003).

The myriad dependencies among species in ecosystems mean that we cannot examine single species in isolation when devising management to protect them or restore their populations to their original levels. Yet, using a conceptualization that navigates through the direct and indirect lines of dependency among species prior to any management action may enable us to anticipate the emergence of unintentional outcomes and ensure we mitigate these outcomes before they happen.

Thus a chronic human-altered environmental factor has the potential to transform the species make-up of large ecosystems by decoupling important direct and indirect linkages among species.

8

Ecosystem Services of Biodiversity

THE BOREAL FOREST, GIVEN ITS CIRCUMPOLAR DISTRIBUTION (FIGURE 3.2), is among the most expansive regions of the globe. As a consequence, it has enormous economic value because it is an important source of timber and pulpwood. Boreal ecosystems are therefore routinely subject to large-scale forest harvesting, often in the form of clear-cut harvesting. A limiting factor in long-term sustainability of the forest industry is the boreal forest ecosystem's capacity to regenerate following harvesting. In particular, boreal forest ecosystems contain mixtures of hardwood aspen (*Populus tremuloides*) that are sought by the pulp, paper, and plywood industries and softwoods like white spruce (*Picea glauca*) that are sought largely by the lumber industry. Historically, attempts to regenerate such mixed woods after clear-cut harvesting have largely met with failure because in many instances only aspen, a dominant competitor species, proliferates and suppresses regeneration of spruce. This gives rise to vast aspen monocultures that risks putting the softwood lumber industry out of business.

Regenerating the mixed wood forest is a desired goal of the forest industry. But current forest management practices can make regeneration an expensive enterprise. It first involves intensive postharvesting site preparation with heavy machinery to create conditions that discourage rapid aspen growth. This is followed by intensive replanting of spruce seedlings, plants that are reared in tree nurseries. In some cases, the return on investment can be marginal at best when these management costs are discounted for net present value over the fifty to sixty year period required to regenerate the forest. One solution to make forest management more cost-effective is to dispense with the expensive site preparation and instead enlist biodiversity and natural ecological processes to carry out the management (Schmitz 2005).

Traditional forest management implicitly eschews biodiversity (specifically plant species diversity broadly and trophic linkages) by intensively managing for a few economically valuable timber species, and discouraging through hunting the presence of large herbivores such as white-tailed deer and moose that are viewed as damaging to regeneration (Schmitz 2005). An alternative view, which may be more compatible with the whole-ecosystem perspective outlined in chapter 1, recognizes that plant species such as aspen and spruce represent dominant and subordinate competitors whose dynamics might be mediated by herbivores (e.g., figure 8.1, cf. figure 2.2c). Deer and moose preferentially feed on aspen because it contains more nutrition than spruce. This leads to the hypothesis that these herbivores might tip the competitive dynamics in favor of mixed wood regeneration. This hypothesis was tested experimentally in a boreal forest site in northwestern Saskatchewan, Canada (Schmitz 2005).

The experiment comprised a systematic comparison of two management treatments: traditional, labor intensive site preparation followed by replanting spruce seedlings, against treatments in which harvested sites were left untouched, including letting natural spruce seedlings grow. Herbivory by deer and moose was further manipulated by allowing herbivores free access to half of both treatment sites and excluding them from the other half. The experiment revealed that deer and moose suppressed aspen regeneration thereby releasing spruce regeneration. The regeneration rate of both tree species exceeded that resulting from traditional management. Moreover, herbivore mediation of aspen-spruce competition resulted in increased diversity of herb and woody plant species that are part of the forest understory (Schmitz 2005).

In the experiment, balanced regeneration of aspen and spruce was best achieved by encouraging greater biodiversity (attracting large herbivores) and enlisting ecological processes (consumer-mediated competition). This case study illustrates that the species within this ecosystem offer an important service: a cost-effective, ecologically compatible way to sustain boreal forest productivity by preserving the trophic structure of the ecosystem.

The recognition that biodiversity offers humankind important services has led to a sea change in thinking about the linkages among biodiversity, ecosystem function, and the economic value of services arising from those functions (Daily 1997). The annual net value of these services is

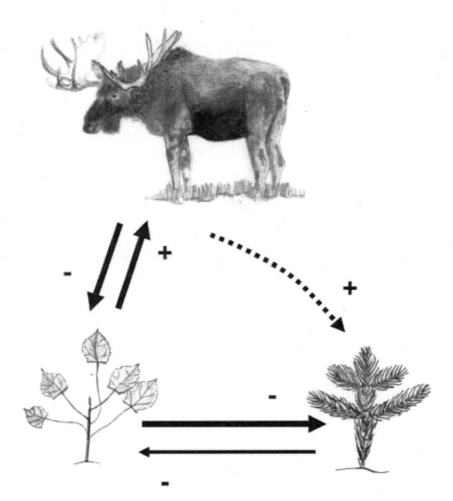

Figure 8.1. Food web diagram illustrating how biodiversity can provide an important ecosystem service, in this case, cost-effective, ecologically compatible forest regeneration. Clear-cut harvesting of boreal forest often favors regeneration of aspen (plant on left). This is because aspen competes with spruce (plant on right), as depicted by the solid arrows between the plants, but aspen is the superior competitor, as depicted by arrow thickness. Mitigating this using traditional forestry practices involves costly heavy machinery that makes site conditions less favorable for aspen growth followed by planting nursery-grown spruce seedlings. An alternative, less costly approach is to recognize that herbivores such as moose prefer to eat aspen, depicted by the consumer-resource (+/−) arrows between moose and aspen. In this case, moose indirectly benefit spruce, depicted by dotted arrow, by suppressing aspen growth and allowing the release of naturally growing spruce seedlings. This represents an example in which enlisting species diversity in several trophic levels of an ecosystem to cause a keystone predation effect can be an effective management tool.

considerable: it has been assessed in the billions of U.S. dollars (Costanza et al. 1997; Daily 1997).

Consideration of ecosystem services fall into two broad categories. The first category—material goods—subsumes contributions with tangible financial value such as new, improved foods, plant based pharmaceuticals, germ plasm infusion for agriculture, raw materials for industry, bioenergy, and so on. These goods can be traded on markets and the markets set values directly through supply and demand pricing. The second category—functions—typically do not have a direct marketable value because they cannot be easily sold. Functions contribute toward economic and financial well-being by sustaining components of ecosystems on which major economies depend (e.g., soil production for farming, clean drinking water, etc.); or by creating opportunities to reduce production costs within those economies (e.g., the above case of herbivore-mediated boreal forest regeneration). In the case of nonmarketable services, the linkages between biodiversity and functions are often less immediate or less direct than they are for material goods and so the linkages require some elaboration—the purpose of this chapter.

> The recognition that biodiversity offers humankind important services has led to a sea change in thinking about the linkages among biodiversity, ecosystem function, and the economic value of services arising from those functions.

Ecosystem functions are both diverse and ubiquitous. The nature and level of the function varies with the make-up and diversity of species in an ecosystem. Ecologists have identified upward of nine classes of function ranging from sustaining ecological cycles (e.g., carbon and nitrogen) to maintaining long-term sustainability of ecosystems (Myers 1996). Below, I highlight specific examples from five categories that have been the focus of experimental research.

- Ecosystem stability and resilience
- Biomass production and minimizing production costs
- Crop pollination
- Pest control
- Resisting invasive species

Diversity Begets Ecosystem Stability

Ecologists have been long been concerned with understanding how ecosystems withstand being buffeted by natural and manmade disturbances. One of the earliest formal explanations invoking a critical role for diversity in stabilizing long-term ecosystem function was developed by Robert MacArthur (1955). It is now known as the Diversity-Stability Hypothesis. MacArthur derived his explanation after wondering why it was that in some ecosystems the abundances of most species changed little in the face of abnormal changes in the abundance of one species in that system, whereas in other ecosystems, species abundances fluctuated widely. The former would be called a stable system and the latter an unstable one (Box 8.1). MacArthur (1955) argued that the stability difference was tied directly to the pattern of interconnectedness or food web linkages among the species in the system.

To understand what MacArthur was driving at, imagine two systems with identical numbers of plant species (P), herbivore species (H), and carnivore species (C) organized into food webs (figure 8.2). Food webs I, II, and III differ only in the way the species are linked together. Yet, these simple differences can have profound implications for the ability of these systems to buffer the effects of a disturbance. For example, any disturbance that lowers the abundance of plant species 1 (P_1) will be felt more strongly in food web I than in food web II or III. The reason is that in food web I, herbivore species 1's (H_1) livelihood is directly tied to the abundance of plant species 1 and the carnivore species depend wholly (C_1) or partly (C_2) on the abundance herbivore species 1. Thus, fluctuations in the abundance of plant species 1 will reverberate right up the food chain and cause significant fluctuations in the abundance of herbivore species 1 and the two carnivore species. In food web II, the effects of fluctuations in plant species 1 will be buffered because herbivore species 1 now has an alternative resource that can help meet shortfalls in its resource supply. In other words, food web II has a greater diversity of species interconnections than food web I. This interconnectedness increases the diversity of resources that higher trophic levels can draw upon to mitigate shortfalls arising from fluctuations in the abundance of any single resource species. By extrapolation, the more interconnected species are in a food web (cf. food web I versus III) the more stable are the systems (MacArthur 1955). This is because the greater *diversity* in the lines of dependency (feeding linkages) among species in highly interconnected food webs allows species higher up in the food chain to

---

Box 8.1 Types of Ecosystem Stability

There are three ways that ecosystem stability may be measured:

Coefficient of Variation—Species abundances (population numbers or total biomass) may fluctuate over time due to disturbances. The degree of fluctuation can be measured statistically by scaling the magnitude of fluctuation (measured as the standard error about a mean abundance) relative to mean abundance. Smaller coefficients of variation indicate smaller fluctuations and hence imply greater stability.

Resistance—Species have some capacity to withstand a disturbance. Resistance quantifies that capacity. A small change in species abundance due to some disturbance implies greater resistance than does a larger change.

Resilience—Species abundances may be changed by disturbances. But, species also have the capacity to recover their abundances after a disturbance. The rate at which a species or an ecosystem recovers is a measure of resilience. Species or ecosystems that recover quickly are more resilient than those that recover more slowly.

---

compensate for species losses lower down. More diverse systems tend to be more stable than less diverse systems.

The Diversity-Stability Hypothesis was tested using a large-scale field experiment at the Cedar Creek Long-Term Ecological Research site in Minnesota, United States (Tilman 1996). In this system, plant species are consumers of soil nutrients in a food web context similar to that envisioned by MacArthur. The experiment comprised different numbers of plant species (from two to twenty)—and hence different numbers of consumer-resource linkages with soil nutrients—planted out among 207 field plots. Tilman measured the degree of fluctuation in total plant biomass in a plot from one year to the next. In this case, stability was measured using the coefficient of variation (CV) in plant biomass (see Box 8.1). The experiment revealed that total plant biomass in plots with greater plant diversity tended to fluctuate less (have lower CVs) than biomass in plots with lower diversity.

In many parts of the world, periodic drought is an important stressor on ecosystems. Frank and McNaughton (1991) evaluated how such stress affected the stability of a grassland ecosystem by tracking changes in plant species composition before and after the 1988 drought in Yellowstone National Park, United States. Within three different locations in the park they

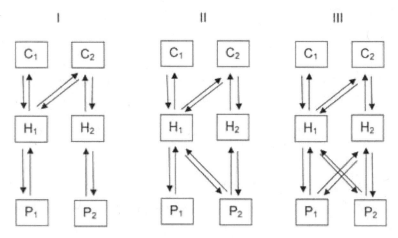

Figure 8.2. Food web diagrams in which the composition of plant (P) herbivore (H) and carnivore (C) species remains unchanged but the degree of interconnectedness in lines of dependency varies. The systems range from low diversity (low degree of inter-connectedness) as depicted in Food Web I to high diversity (high degree of intercon-nectedness) as depicted in Food Web II. In Food Web I species tend to be specialized in their use of prey resources and hence have a unique role to play in the ecosystem that complements the other species. The uniqueness of the role diminishes (i.e., functional redundancy increases) as the degree of interconnectedness increases.

documented that drought-induced species composition change was posi-tively related to grassland plant species diversity (measured as species rich-ness), consistent with expectations of the Diversity-Stability Hypothesis. Although stability was measured in terms of the degree of fluctuation in plant species composition, the study provides insight into another element of diversity known as *resistance* (see Box 8.1). Frank and McNaughton ef-fectively showed that species-rich systems were more resistant (i.e., changed less) to the disturbance than systems with lower diversity.

A review of the broad ecological evidence (McCann 2000) revealed that greater species diversity is related to greater ecosystem stability. The evidence further indicates, however, that species diversity per se does not drive this relationship. Rather, ecosystem stability depends upon the nature of the trophic linkages among species and the relative strength of the inter-actions between species. Specifically, the most stable systems are those in which there are a few strongly interacting species and many weakly interacting species. This asymmetry creates the necessary conditions for the many weakly interacting species to counterbalance the destabilizing effects of a few strongly interacting species (McCann 2000). Thus, weakly interacting species that individually might contribute minimally to a specific ecosystem service may collectively provide an essential functional role in ecosystems.

Ecosystem stability thus depends upon the nature of the trophic linkages among species and the relative strength of the interactions between species.

Although stability was measured in terms of the degree of fluctuation in plant species composition, the study provides insight into another element of diversity known as *resistance*. Frank and McNaughton effectively showed that species-rich systems were more resistant (i.e., changed less) to the disturbance than systems with lower diversity.

Instability brought about by fluctuations in species abundances or biomass also has important financial implications for natural resource-based economies (Armsworth and Roughgarden 2003). Instability increases the risk that an exploited resource will not replenish itself because production of species biomass becomes highly variable. There may even be complete loss of long-term sustainability if the feeding linkages among the harvested species and all others in the system are additionally disrupted by habitat fragmentation as described in chapter 6. Essentially, the Diversity-Stability Hypothesis proposes that maintaining diversity is a way to ensure against fluctuations that cause declining ecosystem function or performance. Thus, biodiversity may provide insurance against loss of ecosystem function (Yachi and Loreau 1999).

Rivets and Redundants

The nature of the feeding linkages among the species in figure 8.2 also determines the functional role of species in the ecosystem. This may have added implications for ecosystem stability, or rather resistance to loss of species. For example, food web I depicts herbivore species 1 and 2 (i.e., H_1 and H_2) as specialists on their respective plant resources (P_1 and P_2); food web III depicts them both as being generalist and completely overlapping in their resource use. In essence, the herbivores in food web I have a unique functional role in the ecosystem whereas in food web III they are completely functionally redundant to one another. The possibility for different functional roles has led ecologists to develop various hypotheses about the link between species functional diversity and ecosystem function.

When species tend to be specialized, as in food web I, they each have a specific "job" in the ecosystem. The emerging level of an ecosystem service, say rate of organic material decomposition for soil creation, rate of nitrogen fertilizer cycling, or rate of plant production, then increases incrementally with each new specialist species that is part of the ecosystem. That is, there is functional complementarity among species. This kind of scenario is the basis for the "Rivet Hypothesis" of species diversity and ecosystem function (Ehrlich and Ehrlich 1981). The Rivet Hypothesis likens species in an ecosystem to rivets on an airplane wing. You can lose a few rivets with limited consequences. If you lose too many, however, the integrity of the whole wing (ecosystem) is compromised. The wing (ecosystem) then loses its ability to maintain critical function and breaks apart. Alternatively, when species are so generalized that they overlap completely, as in food web III, they are able to back-up one another's roles. That is to say they are functionally redundant. The Functional Redundancy Hypothesis of species diversity and ecosystem function (Walker 1992) envisions classes of species doing different kinds of jobs. But, within a class, say grazing herbivores, several species may be doing the same job. In this case, loss of one or a few species from that class does not jeopardize ecosystem function because other species can back up the lost species. The burden of current evidence, derived from a synthesis of fifteen field experiments involving plant communities across the globe, suggests that species tend to play a complementary role to each other (i.e, they tend to be rivets) rather than have a functionally redundant role (Schmid et al. 2001).

> The Rivet Hypothesis likens species in an ecosystem to rivets on an airplane wing. You can lose a few rivets with limited consequences. If you lose too many, however, the integrity of the whole wing (ecosystem) is compromised.

Diversity-Productivity Relations

Much of life on earth is sustained by the fact that green plants use sunlight to stimulate a physiological process—photosynthesis—that converts carbon dioxide and water into sugars and oxygen. The sugars especially are the building blocks that lead to plant structure (roots, shoots, and leaves) and, essentially, a green world. This green biomass, and the food energy contained within it, supports herbivore species higher up in the food chain; herbivores in turn are energy packets that support carnivore species further up the food chain (see figure 8.2). It is easy to imagine, then, that increased production of plant tissue available to herbivores can lead to more herbivores and more carnivores. Plant production is probably the most fundamental ecosystem function supporting life on earth. It also sustains natural resource economies. Without sustained production or yield of grassland biomass, it would be difficult to support a cattle industry. Sustainable logging relies heavily on reliable production of plant biomass in the form of stems (timber). Ecological research is now showing that species diversity contributes toward the amount of plant biomass that is produced in ecosystems.

The most notable example that species diversity plays a role in plant biomass production comes from a systematic multi-site comparative study, called the BIODEPTH (**BIO**diversity and **E**cosystem **P**rocesses in **T**errestrial **H**erbaceous systems) project (Hector et al. 1999). The BIODEPTH project aimed to evaluate not only how plant species diversity contributed toward plant production but also to see if this outcome was repeatable (a hallmark of scientifically reliable insight) by conducting identical, simultaneous experiment protocols in grassland field sites within eight different European countries. The experiment manipulated both plant species richness in experimental plots and functional group richness (e.g., grasses, nitrogen fixing herbs, etc.). Although the exact level of plant production varied among sites owing to properties of the sites themselves (such as soil fertility and moisture and the plant species present at a location), there was nevertheless a strong signal among all of the plots. As plant species richness and plant functional group richness increased, so did plant productivity (Hector et al. 1999). The outcome arose largely because plant species had complementary effects on productivity. That is, species tended to be rivets rather than redundants.

Crop Pollination

Some of the most tightly coevolved relationships in nature involve plants and the animal species that pollinate them. Plants have remarkable varieties of adaptations such as flower shape and color or scent and nectar that are used to attract specific pollinator species. In some cases, there are single or very few animal species that pollinate a particular species of plant making plant-pollinator associations an important source of biodiversity globally.

Bees, in particular, pollinate one or more cultivars of a majority of the world's 1,500 crop species and thus are essential for an estimated 15 to 30 percent of world food production (Kremen et al. 2002). Farmers have relied on such relationships for millennia and to foster efficiency have cultivated extensively the European honey bee (*Apis mellifera*) within crop fields and orchards. This pollination function is now becoming compromised, however, due to diseases and poisoning from insecticides (Kremen et al. 2002). This decline in beekeeping, combined with increasing demand for the service is making cultivation of European honey bees expensive. One solution is to look to natural, related species of pollinators to fill in their ecosystem role.

Maintaining habitat for a diversity of native bees is important for crop pollination because the abundances of individual species fluctuate from one year to the next or they differ in abundance across a landscape. Indeed, up to twenty species of native bees were necessary to ensure sufficient pollination function in any one year (Kremen et al. 2002). Also, different bee species are differentially effective as pollinators for different crops. In essence, there is an element of both functional redundancy (bee species covering for each other among years and in different locations) and functional diversity or rivets (different groups of bee species effective on different plants) in this pollinator community. Managing for bee diversity could therefore meet the pollination requirements of a great number of crops and provide insurance against shortages of domesticated honeybees and specific native pollinator species (Kremen et al. 2002).

> Managing for bee diversity could therefore meet the pollination requirements of a great number of crops and provide insurance against shortages of domesticated honeybees and specific native pollinator species.

The pollination service offered by native bees can, however, vary with land management practices. To understand quantitatively just how land use affected pollination function, Kremen and colleagues undertook a systematic study to document how diversity in the native bee species community affected the rate of crop pollination on farms in the Central Valley of California. They found that amount of crop pollination depended on how close a farm was to natural bee habitat and on the type of farm itself (organic versus conventional). Native bees had the greatest effect on organic farms near natural habitat. All other farms (e.g., organic far from natural habitat, conventional near or far from natural habitat), however, experienced greatly reduced diversity and abundance of native bees, resulting in some cases in failure of the native bees alone to pollinate crops. This implies that farmers need to consider managing farmland for both natural pollinator habitat as well as crop lands.

Pest Control

Herbivorous insect species are a leading cause of crop plant damage and accordingly annual loss of crop production. Suppression of such insect pests by their natural enemies has been identified as a potentially important ecosystem service of biodiversity (Cardinale et al. 2003).

A case in point involves one of the most abundant pests of alfalfa crops, the pea aphid, (*Acyrthosiphon pisum*). Three of the most important natural enemies of pea aphids are the ladybird beetle (*Harmonia axyridis*), the damsel bug (*Nabis sp.*), and the parasitic wasp (*Aphidius ervi*). Cardinale et al. (2003) conducted an experiment in alfalfa fields in Wisconsin, United States, to compare the effectiveness of the three natural enemies individually (low diversity) and in combination (high diversity) in controlling aphid abundance. They discovered that the abundance of the pea aphid was suppressed more under high diversity than low diversity conditions. Aphid reduction translated into higher alfalfa production.

Moreover, the degree to which pea aphids were suppressed in the high diversity condition was greater than one would expect simply from summing the impact of each enemy species alone. This nonadditive effect arose in part because different predator species were functionally complementary in their effects on different age classes of the aphid, and hence set up a predation gauntlet. The parasitic wasp deposits eggs into the body of young aphids. The eggs hatch and the wasp larvae consume the internal tissues of the young aphids. The ladybird beetle and the damsel bug actively hunt older aphids. Cardinale and colleagues, however, caution against ascribing universality of this finding for all natural enemy species. In some cases, natural enemies may overlap in ecological roles (functionally redundant) and may even attack each other as well as their shared prey (called an intraguild effect) in which case the effectiveness of the diverse natural enemy assemblage is compromised. The underlying message is that natural enemy diversity is effective provided one is strategic about predator species selection. Predator species chosen for a particular biological control problem should be ones that complement each other in their effects on a pest species. But, to allow options for future biological control, it is imperative that a portfolio of predator species (i.e., predator species diversity) from which to choose is actively maintained on ecological landscapes.

Invasion Resistance

As explained in chapter 7, biotic invasions occur when species are transported beyond their current range distribution into conditions in which their descendants thrive. The likelihood that a species can invade depends in good part on biophysical features of the environment. For example, the success of invasions may depend upon the availability of bare areas, requisite soil fertility, and suitable climate (Kennedy et al. 2002). These biophysical conditions for invasion may be mitigated by the presence of native species that can regulate the competitive environment the invader faces. So, it stands to reason that the likelihood of successful invasion may decline as the number of native species with which the invader must compete increases. In essence, biodiversity should lead to ecosystem stability by resisting species invasions and associated decline in ecosystem function (see Box 8.1; Kennedy et al. 2002).

This idea was tested using a large-scale field experiment at the Cedar Creek Long-Term Ecological Research site in Minnesota, United States (Kennedy et al. 2002). The experimental system consisted of 147 different experimental plots in which resident plant diversity (species richness) was initially varied from one, two, four, six, eight, twelve, and twenty-four grassland species drawn at random from a pool of twenty-four species. Invaders were allowed to enter the experimental area from the surrounding weedy fields through natural dispersal processes. The experiment revealed that species richness had a strong suppressing effect on invader establishment success (there were fewer invaders in high diversity plots than in low diversity plots) and on the ability of invaders to proliferate (invaders had smaller maximum size in high diversity plots than in low diversity plots). The study suggests that loss of biodiversity may exacerbate likelihood of invasion; it can diminish resistance against invasion once other factors that insulate habitats from invasion, such as natural geographical barriers, are compromised by human transport of nonnative species.

These examples illustrate that biodiversity can have important functions in ecosystems, the value of which may go well beyond that measured in terms of material goods. Conserving such functions should not be viewed as competing with economic activity, but rather as essential and consistent with a pursuit of human well being (Myers 1996).

9

Protecting Biological Diversity and Ecosystem Function

AN IMPORTANT LESSON FROM CHAPTER 6 IS THAT CONSERVATION OF SPECIES requires preservation of natural habitat. Upsetting the integrity of natural habitats through fragmentation, exploitation, or conversion to other land uses can trigger a cascade of ecological changes that sooner or later lead to species extinctions. Indeed, it is estimated that if the current pace of human-caused global habitat loss continues, many habitats and associated species will be completely eliminated by 2080 (Sinclair et al. 1995). More than ever in the history of the planet, humankind is at a critical juncture in the way it chooses to interact with the natural world.

The practical reality is that the need to support a burgeoning global population demands that natural lands be increasingly exploited for their raw materials or converted into living space and agricultural production. The attendant consequence is that there is altogether less space to support the rest of planet's living diversity, and so humankind is forced to become strategic about what species it actively chooses to protect. Our ability to do this is constrained by limited funding. To reconcile this trade-off, we have to look for ways to conserve the most species per dollar spent.

One expedient strategy is to identify areas that support high concentrations of species—"biodiversity hotspots"—and devote efforts at protecting those areas (Myers et al. 2000). The rationale for adopting such a strategy is that biodiversity hot

The practical reality is that the need to support a burgeoning global population demands that natural lands be increasingly exploited for their raw materials or converted into living space and agricultural production.

spots contain exceptional numbers of endemic species (species only native to those locations and that have a long evolutionary history there) that are, at the same time, facing rapid loss of their native habitat. A biodiversity hotspot strategy creates the opportunity for conservation planners to target, from a global perspective, a comparatively small fraction of land that supports a disproportionately high share of the world's species at risk.

> A biodiversity hotspot strategy creates the opportunity for conservation planners to target, from a global perspective, a comparatively small fraction of land that supports a disproportionately high share of the world's species at risk.

To this end, conservationists have identified twenty-five global biodiversity hotspots. Many of these hotspots are situated within the equatorial regions of South and Central America, Africa, and Asia (Myers et al. 2000). These areas are the sole remaining habitats of 44 percent of the earth's plant species, 28 percent of all bird species, 30 percent of all mammal species, and 54 percent of all amphibian species (Myers et al. 2000). The hotspot areas originally covered 17 million square kilometers (an area of about twice the size of Canada). Habitat loss from exploitation and land conversion has reduced that area to 2 million square kilometers or about 0.6 percent of the earth's total surface. It is speculated that if the current pace of habitat loss proceeds virtually unchecked, between one-third and two-thirds of all species in the hotspot areas would disappear within the foreseeable future (Myers et al. 2000). Accordingly, protecting the twenty-five hotspots might stem this decline significantly. At the same time, saving species within hotspots putatively resolves the trade-off because it allows for human development of the remaining global land base (Myers et al. 2000).

Conservation Tools

There are a number of approaches that one can take to protect biological diversity (Box 9.1). These approaches differ in the degree to which they trade-off human land-use and protection.

Parks and Protected Areas

Protecting species within hotspot areas involves circumscribing the hotspot areas with a fixed political boundary and legally designating the area within those boundaries as national parks or protected areas (Box 9.1). To safeguard species and natural habitats, access to parks would be legally restricted or

---

Box 9.1 Conservation Tools to Protect Biodiversity

National Parks and Preserves—The intended goal is to protect species, large areas of scenic, natural beauty, and natural processes in as undisturbed a state as possible for scientific, educational, or recreational use. Park use and policy is regulated at the national level.

Indigenous Reserves or Biosphere Reserves—recognize the land tenure rights of indigenous peoples. It provides the opportunity to maintain their traditional subsistence livelihoods on the land. Such reserves represent samples of landscapes from long-established land use patterns.

Extraction Reserves—allow local economies to develop through local commercial enterprises (e.g., rubber tapping, rearing Brazil nuts, etc.). Such reserves allow people to continue with their traditional economic way of life. This can lead to local sustainable economic development. The success of such programs rests on national government guarantees that land conversion to other enterprises (e.g., farming, ranching) will not take place.

---

precluded and enforced by park wardens. Therefore, people who have historically resided within the hot spot areas would be displaced and forced to live with others in a common land matrix outside the park boundaries. Such a strategy is one way to reconcile the conflicting needs of conservation and economic development simultaneously on the land base. Both activities are permitted, just in different locations.

But, most hotspots lie within the heart of developing countries with extensive poverty. Human population density in hotspot areas is on average seventy-three persons per square kilometer, a number that is 73 percent higher than the world average of forty-two persons per square kilometer. Thus, displacing indigenous peoples from the hotspot areas and concentrating them with others in urban areas could increases the potential for huge conflicts because such a conservation strategy can pit conservation interests against human livelihoods (Schwartzman et al. 2000). Indeed, concentrating people in the land matrix between the parks can lead to serious overexploitation of resources. Once resources are exhausted, displaced people look to remaining vestiges of resources. These remaining resources often lie within parks. For example, on-the-ground monitoring (Curran et al. 2004) has shown that within sixteen years (1985 to 2001), Kalimantan's *pro-*

tected lowland forests declined by more than 56 percent. Even uninhabited frontier parks like Gunung Palung National Park in West Kalimantan were almost completely logged to meet timber demands (Curran et al. 2004).

Because of their fixed political boundaries, parks and protected areas strategies also have the potential to predestine systems to other forms of extinction debt because of a failure to consider landscape dynamics (Carroll et al. 2004). Specifically, development of the land base between parks increasingly isolates the parks themselves. From a landscape dynamic perspective, parks effectively represent isolated fragments of the original, larger habitat. Our consideration of species-area relationships in chapter 6 taught us that fragments can support only a fraction of the species supported by intact habitat. Moreover, development within the land matrix between the fragments makes it difficult for species to migrate and recolonize those fragments in which species have gone extinct. That is, there is no possibility to "rescue" species that go extinct within the park. Isolating parks thus can precipitate chronic extinctions (Newmark 1984).

In addition, parks may fall short if there is a failure to align legal and biotic boundaries (Newmark 1985; Caroll et al. 2004). A case example is Algonquin Park in central Ontario, Canada, a place that is highly revered for its population of wolves, moose, and white-tailed deer. The name itself harkens back to an age when native Indians occupied the landscape; hence the park symbolizes primordial wilderness. The region was the inspiration for an important impressionistic art movement—the Canadian Group of Seven—that portrayed natural landscapes in their raw glory and thereby forged an indelible and important Canadian wilderness identity. In the 1800s, loggers harvested the vast stands of white pine (*Pinus strobus*) trees in this area to feed the high demand of an expanding British economy. Algonquin Park was created in 1893 to establish a wildlife sanctuary and by excluding agriculture on harvested land to protect the headwaters of the five major rivers that flow from the park. But, humans have been permitted to transform the landscape around the park into agricultural lands. This is where the problem begins.

During fall, white-tailed deer migrate out of the park to over-winter in the surrounding agricultural land matrix. They return to the park in spring and spend the summer there (Forbes and Theberge 1996). Wolves naturally follow their prey's seasonal migration. But, wolves are shot with impunity once outside the park boundary because humans believe that they are a threat to their livestock and competitors for deer, which they also hunt

(Forbes and Theberge 1996). Over a period of thirteen years, fully 56 percent of all Algonquin Park wolves were killed by humans after they migrated out of the park; and 70 percent of those killed were potentially reproductive individuals (Forbes and Theberge 1996). The wolves, (a species that reproduce at a rate of 15 to 25 percent per year), do not have the reproductive capacity to compensate for the high mortality and so could be at risk of extinction.

The human-wolf conflict arises for several reasons. Failure to consider landscape dynamics means that the park is too small to contain a viable year-round population that depends on a reliable prey base. In other words, the park was established to protect a collection of species rather than to protect an ecosystem in which there are myriad lines of dependency among species. Consequently, the park boundary does not contain the complete range of critical habitats used by various species that depend on each other. Also, regulations for wolf protection differ inside and outside the park. Protecting predators leaving the park is viewed as encroaching on the rights and safety of surrounding residents. Clearly, the problem isn't the park by itself, but what happens within the land matrix surrounding the park. A similar plight occurs for large carnivores in and around the national parks within the Rocky Mountain region of North America (Carroll et al. 2004) and for lions in the Gir Forest of India (Saberwal et al. 1994). As we learned in chapters 1 and 6, the loss of such apex predators from ecosystems can precipitate loss of species diversity in those systems. Thus, loss of just one species—but one with a critical functional role in an ecosystem—can confound well-intentioned efforts to conserve concentrations of biodiversity (Soulé et al. 2005).

Ecologists are increasingly realizing that they need to solve the extinction debt problem not by establishing more parks but by recognizing that ecosystems often extend beyond park boundaries. Thus, one must be strategic about linking conservation efforts within parks to the land base outside of them. One strategy, in particular, is to consider enfranchising local peoples in conservation and allowing different forms of human land use both within and around parks. Many different strategies have been proposed in two broad categories: indigenous reserves and extraction reserves (Box 9.1).

An indigenous reserve strategy recognizes and respects this history and thus protects a way of life as well as the species diversity that is part of that landscape.

Indigenous and Extraction Reserves

Indigenous reserves recognize historical land tenure rights of local indigenous people. For example, many agrarian societies have transformed land bases (e.g., terracing mountain slopes) in order to adapt their farming to the vagaries of the landscape within which they work and live. Such societies have a long history on the land base. An indigenous reserve strategy recognizes and respects this history and thus protects a way of life as well as the species diversity that is part of that landscape.

Extraction reserves go a step further by supporting enterprises that are sustained by ecosystem services unique to a particular region in which there is also a need to protect biological diversity. For example, Brazilian tropical rainforest ecosystems contain tree species that produce rubbery latex and others that produce the famous Brazil nuts. These resources support important subsistence economies and can lead to sustainable economic development. In creating extraction reserves, the Brazilian government forges an agreement with local communities to provide for the sustainable use of the forest. At the same time, local communities must guarantee that they will protect the natural integrity of the forest ecosystem. In return the communities would have control over the products from the forest. In other words, control over conservation is ceded by national governments to local communities.

There is divided opinion about the viability of extraction reserves. Some argue that national control over conservation is more stable in the long term than local control because humans, being selfish, will overexploit resources to their own economic gain (Terborgh 2000). It is believed that local stewardship will only work when population densities are low. When resources become stretched by higher densities, humans will extirpate biodiversity for their own gain, or trigger events that lead to collapse of biodiversity. It is also argued that as the standard of living rises because of economic development, local communities will look toward further exploitation of resources to fuel their economic growth, especially when wealth generation enables them to replace aboriginal subsistence technology with modern technology (Terborgh 2000).

> In creating extraction reserves, the Brazilian government forges an agreement with local communities to provide for the sustainable use of the forest.

Advocates for local control of conservation argue that national parks alone will be of insufficient size to protect natural areas in perpetuity (Schwartzman et al. 2000). Moreover, local people protect larger areas of land than are now protected in most parks. In many cases, humans have lived in these regions for centuries and have not wrought mass destruction during that time. Finally, the practical reality is that local people are potent political actors and an important environmental constituency that can determine the success or failure of national parks.

Graded Protection

One compromise strategy may be to design a reserve system on landscapes that allows different degrees of use. For example, one might completely protect a core region that is surrounded by concentric rings (buffer zones) that allow graduated degrees of resource use and land development. Such a strategy also balances the conflict between protection and development, but it achieves this balance on the same parcel of the land base. Allowing conservation strategies that produce economic benefits from goods and functions—ecosystem services—provided by biodiversity within reserve regions represents a new shift in conservation paradigm. Pure protection strategies are giving way to those fostering sustainability through economic incentives to protect nature.

Dynamic Landscapes

Parks and protected areas are typically created to protect areas that are unique or representative of a particular ecosystem type (Box 9.1). As such, this conservation tool only conserves a static entity: diversity within a specific local area or alpha (α) diversity. However, the make-up of a particular local community of species may be shaped by sorting processes that operate on a larger, landscape scale—called a metacommunity process.

Metacommunity Dynamics

The metacommunity concept is similar to the metapopulation concept discussed in chapter 6 except extinctions are assumed to be determined by species interactions in local habitats. It represents a way to understand

> The make-up of a particular local community of species may be shaped by sorting processes that operate on a larger, landscape scale—called a metacommunity process.

how species come to occupy different regions of a landscape via two ecological processes: migration across landscapes, and interactions such as competition and predation within local areas (Leibold et al. 2004). The
metacommunity is represented by the set of local species assemblages or
local communities. The constellation of local communities on a landscape
is shaped by the rate of species migrations among locales relative to the
strength and nature of species interactions within a locale.

The metacommunity allows evaluation of another form of diversity,
known as beta (β) diversity. Technically, beta diversity quantifies the rate at
which species compositions turn over or become dissimilar across incre-
mental distances on landscapes. A high turn over rate or high dissimilarity
between adjacent locations implies that the species composition across the
landscape is highly heterogeneous, or is highly diverse.

The implication here is that areas deemed by humans to be representative of a certain environmental state or a local hotspot could very well be
the outcome of landscape-scale processes that produce a local concentration of species rather than, what is usually implicitly assumed,
that they represent local sources of diversity for the landscape. Accordingly, a conservation strategy that creates parks and permits land devel-
opment within the matrix between parks can disrupt the flow of species across landscapes and alter beta diversity. Moreover, such disruption could eventually doom even the hot spot
species pool to extinction. This insight underscores that effective conservation requires thinking about protecting dynamic ecological processes leading to diversity patterns across landscapes, rather than simply identifying and
protecting diversity concentrations locally within landscapes.

> Beta diversity quantifies the rate at which species compositions turnover or become dissimilar across incremental distances on landscapes. A high turnover rate or high dissimilarity between adjacent locations implies that the species composition across the landscape is highly heterogeneous, or is highly diverse.

Temporal Dynamics of Habitats

Habitat, being comprised of living organisms, may not persist indefinitely
in any one location (Sinclair et al. 1997). Dominant vegetation in a location
passes through different stages of a developmental cycle including aging,
dying back, and regenerating. This process is known as ecological succes-

sion. Thus, landscapes may be considered mosaics of habitats that locally fluctuate among successional stages. So, the exact location of a particular developmental stage of habitat will vary in time. So too will the animal species composition as species move across the landscape to follow the changes in habitat. High beta diversity can arise in this case because different animal species associate with these different successional habitat stages.

Thus, a park or protected area strategy that safeguards a fixed parcel of land in one location may doom its species to extinction simply because it ignores the consequences of successional change. One solution then is to actively manage to arrest succession. But arresting succession will require "fighting nature" and history has repeatedly shown us that humankind is doomed to lose at this game. How then do we work with nature? We do this by thinking about habitat and biodiversity conservation in the context of landscape-scale habitat renewal (Sinclair et al. 1995).

Habitat of a specific type in a specific location is lost over time because of the natural ecological processes described above, but the rate of habitat loss can be exacerbated by exploitation. If habitat facing exploitation is preserved at a particular time, such preservation will slow the decay rate. But, it will still decay. So, parks and protected areas are essential because they buy time. But, to maintain a constant availability of habitat, we must actively plan for renewal across the landscape within and outside of protected areas.

The idea of including habitat renewal in conservation derives from the fundamental principles of population processes presented in chapter 4. Namely, a population remains at an equilibrium if birthrate exactly balances death rates. We can extend this simple concept to this case by recognizing that habitat decay is effectively a form of mortality. Thus, habitat of a particular type will be preserved if the habitat decay rate is exactly balanced by habitat renewal. So, the decay rate and the renewal rate determine how much habitat can be saved in perpetuity—a form of sustainable management. But, unlike population birth and death processes that may occur simultaneously within a locale, habitat decay and renewal occur in different locations on a landscape. Moreover, to maintain habitat, we must either replace it before or (at the very least) at the same time as other portions of the habitat are exploited or decay. The implication here is that long-term conservation requires a portfo-

> Habitat of a particular type will be preserved if the habitat decay rate is exactly balanced by habitat renewal.

lio of habitats in different development stages to ensure that habitat of a particular stage is represented on the landscape at any particular time.

Global Climate Change and Reshuffling of Faunas

In chapter 3 we learned that many species are uniquely adapted to live in specific ecosystems on this planet. Those adaptations include tolerance to hot or cold, dry or wet, sun or shade. We also learned that global climate change is being brought about by human activity exacerbating natural greenhouse warming in ways that will alter the earth's climate. The consequence is that habitats and the animal species that depend upon them are expected to undergo major shifts in their geographic locations (Parmesan and Yohe 2003). This may be the Achilles heel of national parks and protected areas that are designed to protect species and their habitats within the confines of fixed political boundaries.

Indeed, an analysis evaluating the effectiveness of eight selected U.S. national parks in protecting mammalian species diversity in the face of global change suggests that they are not likely to meet their mandate of protecting current biodiversity within park boundaries (Burns et al. 2003). Based on current assessments of future climate change, U.S. national parks stand to lose between 0 and 20 percent of their current mammalian species diversity in any one park.

Species losses should come from all mammalian species except Artiodactyla (hoofed mammals). The majority of losses will be for rodent, bat, and carnivore species. But, the parks should also gain other species because of the reshuffling of species across the landscape (Burns et al. 2003). Parks are expected to gain between 11 percent and 92 percent more species relative to current numbers. Species reshuffling is predicted to be dominated by an influx of rodents followed by carnivores and bats. The projected influx of new species is expected to be greatest for parks at more northerly latitudes because, as described in chapter 3, most species are expected either to expand their geographic range or shift their ranges to new geographic locations nearer the poles.

North American parks could realize a substantial gain in mammalian species composition as a consequence of geographic shift in species. But this shift will be of a magnitude unprecedented in recent geological time.

In the balance, all parks could realize a substantial gain in mammalian species composition as a consequence of geographic shift in species. But this shift will be of a magnitude unprecedented in recent geological time. Also, it is assumed that species will reshuffle en masse, in an orderly manner and that the rate of distribution change is commensurate with geographic shifts in habitat. Such comparatively rapid (twenty to fifty years) range adjustments are not entirely out

> Successful conservation increasingly requires that we accommodate both human needs for sustainable livelihoods and the need to protect the *dynamics* of ecological systems that play themselves out across landscapes.

of the question for mammals (Burns et al. 2003). Nevertheless, all of this assumes that species will be free to migrate in through the land matrix between the parks, an assumption that, as discussed in chapter 3, may not be realistic for many species given the kinds of land development taking place.

There may be further ecological repercussions. As shifting species forge new ecological relationships with each other and with current park species, the character of species interactions and fundamental ecosystem processes stand to become transformed (Burns et al. 2003). For example, an influx of new species may alter existing competitive interactions and influence trophic dynamics as predator-prey interactions change.

These considerations of landscape-scale dynamics means that classic conservation approaches, which rely almost exclusively on the establishment of parks and protected areas, will require some serious rethinking. Successful conservation increasingly requires that we accommodate both human needs for sustainable livelihoods and the need to protect the *dynamics* of ecological systems that play themselves out across landscapes. So, parks and protected areas need to be viewed as part of a larger portfolio of options. Such portfolios allow for more flexibility because the specific options exercised will depend upon local needs for conservation and development.

10

The Good of a Species: Toward a Science-Based Ecosystem Conservation Ethic

HUMANKIND IS AT A CRITICAL JUNCTURE IN THE WAY IT CHOOSES TO IN-teract with the environment. Rising human population combined with ris-ing demand for resources to sustain or even elevate standards of living means that living space for other species is diminishing. The current sentiment is that we cannot protect everything and so we must be strategic about what we keep and what we choose to let go. To this end, society is increasingly demanding that ecologists explain what species A, or species B, and so on, is good for (Myers 1996). This form of questioning reveals a deeply held ethic that species are largely expendable showpieces; hence it is reasonable to do ecological triage if allocating space or financial resources to conserve species conflicts with human economic progress (Myers 1996).

Aldo Leopold (1953) had a different take on this same issue:

> The outstanding scientific discovery of the twentieth century is not television, or radio, but rather the complexity of the land organism [ecosystem]. . . . The last word in ignorance is the man who says of an animal or plant [species] "What good is it?" If the land mechanism as a whole is good then every part is good, whether we understand it or not. If the biota, in the course of aeons, has built something we like but do not understand, then who but a fool would discard seemingly useless parts? To keep every cog and wheel is the first precaution of intelligent tinkering.

Leopold effectively argued that society needs to change its ethical per-spective about nature from one that views it as a magnificent collection of species for our passing enjoyment to one that recognizes that species play integral roles in maintaining ecological services that make up nature's econ-

126

omy. Thus asking an ecologist to justify the value of a species or a collection of species in an ecosystem is tantamount to asking an economist to justify why we value securities such as stocks or mutual funds. I believe that both are integral components of their respective economies and without them it would be impossible to carry out our day-to-day business of living. But, there are fundamentally different opportunity costs to managing portfolios of stocks versus portfolios of species.

Tinkering within Economies

In a market economy, intelligent tinkering is done by investors who adjust their portfolios with the aim to balance financial return against financial risk. Such a balancing act can be accomplished by keeping those stocks that perform well and letting go those that perform poorly. But, the performance of individual stocks changes over time so the portfolio must be readjusted, including perhaps, buying back stocks that were once performing poorly but are now performing better. Thus, in a market economy, a security that is excluded from a portfolio can be included again later provided that the company offering the stock is still in business. That is, there is opportunity to reverse investment choices. Moreover, one has flexibility to make rapid adjustments by buying and selling stocks in the very short term.

Leopold effectively argued that society needs to change its ethical perspective about nature from one that views it as a magnificent collection of species for our passing enjoyment to one that recognizes that species play integral roles in maintaining ecological services that make up nature's economy.

Intelligent tinkering with an ecological portfolio requires that we appreciate some important differences between market and natural economies. First, and foremost, natural economies obey classical laws of thermodynamics, which means that we cannot "grow" them indefinitely. Species populations, like capital, may accrue for some time through a compound interest process. But with financial capital, the investment interest rate does not change because capital accrues. Unlike capital, species population growth rate eventually does decline because species density causes negative feedbacks on population growth rate. This then constrains ecosystem functioning and production within upper limits. (See chapter 4 for further discussion of these dynamics.) Working within those limits means that sustaining natural economies cannot be about deciding which species in the portfolio should be let go. Rather, it is about recognizing the need to keep all species as part of the portfolio. That is, conservation should be about managing for sustainable function of ecosystems, not about protecting selected species. The goal of management, then, is deciding what abundance of the different species should make up the portfolio in order to achieve some desired end. This reasoning is predicated on the idea that species play complementary rather than strictly redundant roles in determining ecological functions, for which there is some evidence (chapter 8).

Second, natural economies operate on longer timescales than most market economic systems. There isn't a measure of daily performance such as a Dow-Jones or NASDAQ index for natural economies because they do not respond to changes that quickly. As illustrated in chapter 7, the effects of tinkering with an ecological portfolio (e.g., predator species removals to enhance game species populations) can require decades to half centuries to fully manifest themselves in ways that are detectable by scientific measurement. Thus, the consequences of our ecological investment choices often will only be fully realized by our children and grandchildren. This cautions against a strategy of species triage because that inevitably dooms some species to extinction. That is, new companies and stocks can be freely created; new species cannot. Thus, the potential for irreversibility in investment choices created by triage strategies paints our children and grandchildren into a proverbial environmental corner.

> Intelligent tinkering with an ecological portfolio requires that we appreciate some important differences between market and natural economies. First, and foremost, natural economies obey classical laws of thermodynamics, which means that we cannot "grow" them indefinitely. Furthermore, the consequences of our ecological investment choices often will only be fully realized by our children and grandchildren.

I have, nonetheless, offered examples of investment reversibility in a natural economy by highlighting in chapter 7 the case of wolf reintroductions into Yellowstone Park. Restoring the top carnivore led to changes in the distribution and abundance of the wolves' major prey species, elk, and increases in the abundance of tree species that are both food resources for elk and habitats for other animal species. Devising ways to restore ecosystems, to correct what in hindsight may have been imprudent choices, is an important enterprise in ecological science. But reintroducing once-lost species back to their native habitats can be a formidable task because it presumes that the we can draw individuals from somewhere else on the landscape (i.e., the species have not yet gone extinct) and that suitable habitat is available to support the reintroduction.

Species also do not live in splendid isolation of each other. There are direct and indirect lines of dependency that make up the natural economy (chapters 7 and 8). These dependencies can have an important bearing on the success of reintroductions, especially if we have limited insight into how they arose over the eons in the first place. This uncertainty leaves us with the sobering prospect of what Pimm (1991) has termed "Humpty Dumpty effects."

The concept has its roots in the familiar nursery rhyme in which all the parts of a broken Humpty Dumpty are present, but he cannot be put back together because it is unclear how the pieces fit together originally. To elaborate this point, let's consider a hypothetical restoration scenario and use the chaparral food web (figure 6.2) for illustrative purposes. In chapter 6, I explained how habitat fragmentation led to local extinctions of bird species that comprised the chaparral food ecosystem because fragmentation disrupted important lines of dependency among top and middle (meso) predators whose effects cascaded down to bird species causing local bird species extinctions. Suppose that a conservation agency decided that it would restore the chaparral ecosystem in southern California by buying parcels of land to reverse development and reconnect that habitat fragments. Suppose also that the conservation agency could find individuals of the different bird species elsewhere on the landscape and that it decided to reintroduce them en masse in order to hasten ecosystem recovery. This strategy might be successful, but it might not because of an ecological phenomenon known as priority effects. Priority effects mean that species in an ecosystem do not assemble haphazardly but instead follow a strict sequence.

For example, let's suppose a primordial chaparral habitat was comprised only of the roadrunner and its prey: insect species and lizards. Suppose that from time to time the other bird species (the wrens, quail, and gnatcatcher) attempted to become established in the chaparral but they were unsuccessful because roadrunners, which also prey on bird eggs, prevented the other bird species from growing in abundance. Suppose, however, that over time, the roadrunner population built to a sufficient size to support a population of carnivores such as foxes that preyed upon roadrunners. The limitation of roadrunner abundances by foxes would now create opportunity for the other bird species to become established. Let's suppose that this led to the rich diversity of birds in a more modern chaparral ecosystem. Let's suppose now that foxes began to build their own populations by diminishing the abundance of roadrunners. Hungry foxes then might switch to preying on the other bird species. But, suppose also that because foxes became sufficiently abundant so that coyotes could become established by preying on foxes. This would release predation pressure on the bird species. So coyotes now indirectly maintain bird species diversity in the ecosystem. The direct and indirect lines of dependence leading to the maintenance of species diversity and function in this ecosystem became established through a specific order of priority from roadrunner to fox to other bird species to coyote. Any restoration sequence that deviates from the original assembly process, including reintroducing species en masse, could very well lead to failure—a Humpty Dumpty effect.

Some might dismiss all this by saying that in the grand scheme of things it is just a few insignificant little birds and mammal varmints that are found in other places anyway—the triage ethic. But, what if the loss of these putatively insignificant species and varmints lead to ecosystem conditions that fuel more frequent and intensive chaparral wildfires that annually damage expensive real estate? Right now we do not have the scientific knowledge to speak to this issue nor to many other issues concerning the role of humans in natural economies. Resolving this is one of the major challenges in the next generation of ecological science (Lubchenco et al. 1991; Ludwig et al. 2001). On the other hand, the prospect of Humpty Dumpty effects could be eliminated altogether by ensuring that we cause minimal risk to species when developing or exploiting ecosystems. In this respect, it is wise to revisit the insight of Aldo Leopold that was presented above:"If the biota, in the course of aeons, has built something we . . . do not understand, then who but a fool would discard seemingly useless parts? To keep every cog and wheel is the first *precaution* [my emphasis] of intelligent tinkering."

Ecological Science, Uncertainty, and Precaution

Sustaining natural economies, like market economies, involves making decisions in the face of uncertain knowledge. Ecological science tries to diminish this uncertainty through a systematic process of induction, deduction, and experimentation (chapter 2); but ecological science cannot eliminate uncertainty completely. This uncertainty has some important ethical implications for decision-making. Let me illustrate with an example.

Suppose that a developer wished to convert a large tract of forested land into housing and commercial enterprises and engaged a scientific consultant to offer insights about the degree of fragmentation that could be sustained without disrupting the diversity of animal species within the land base. Suppose that the consultant executed an experiment like the one conducted in Amazonia (chapter 6) to test the null hypothesis that habitat fragmentation did not cause loss in species diversity. Let's suppose that unlike the Amazonian case, experimental fragmentation did not cause a change in species diversity. We would then conclude that the evidence is insufficient to reject the null hypothesis and thus permit development to go ahead.

But how reliable is the conclusion from this single test? The answer is: limited. The reason is that we are unsure if the outcome is normal for this kind of ecosystem or a rare event that depended on specific place and time. Thus, a single experimental test of a hypothesis cannot provide a highly reliable answer. The only way to increase reliability of knowledge acquired from ecological experimentation is to repeat the experiment many times. That is why ecologists, when asked to offer scientific advice, often qualify their conclusions with the disclaimer "this needs further study . . . .", meaning, this needs more replication to gain general insight into the likely trend.

Such a disclaimer is a perennial source of frustration for policy makers who seek definitive answers to specific problems. Yet the disclaimer is legitimate because ecologists can never be definite that one factor or variable is the causal driver of a pattern or process. This is because repeated experimental tests of the same hypothesis will never produce identical (invariant) outcomes in the field due to random environmental effects. Consequently, ecologists are only able to present their scientific results in terms of the likelihood or risk that a particular event will happen (Ludwig et al. 2001). Indeed, this is true for any scientific discipline that must deal with confounding effects of random environmental factors encountered in natural environments.

> A single experimental test of a hypothesis cannot provide a highly reliable answer. The only way to increase reliability of knowledge acquired from ecological experimentation is to repeat the experiment many times.

Variability and associated uncertainty introduced by random environmental effects can be dealt with by repeating experiments enough times that we can calculate the odds (or risk) that an experimental treatment (or a management prescription based on it) provides a rare or unusual outcome. For example, if we conducted the same fragmentation experiment twenty times under different environmental conditions and found that it does not alter species diversity (that is, we could not reject the hypothesis) nineteen times out of twenty, then we say that there is a 5 percent chance that habitat fragmentation will collapse bird species diversity. The implication of this calculation for policy is that if the goal is to develop land and safeguard species diversity, then we can claim that there is a 5 percent chance that fragmentation will be harmful. This information is valuable because it allows policy makers to decide whether the benefit of implementing a particular management prescription is worth the risk of failure.

But uncertainties still underlie decision-making, which opens the possibility for making errors. The first kind of error, known as a Type I error, arises when we reject the null hypothesis when the null is in fact true. The likelihood of doing this can be kept marginally small simply by being very stringent about the criterion we use to reject the hypothesis. The norm in ecology is a 5 percent risk of making a Type I error. Criteria for ensuring a small likelihood can be calculated whenever it is possible to quantify the mean and degree of variability in experimental response among different replications—often this can be done with as few as three replications. The second kind of error, known as a Type II error, arises when we accept the null hypothesis when in fact it is false. Type II errors can only be controlled by gaining understanding of typical experimental responses versus comparatively rarer ones. Our confidence in discerning what is typical and what is rare is only boosted by increasing the number of times an experiment is replicated. But, most experiments conducted on scales relevant to ecosystem management are neither easy, nor practical, nor often affordable to replicate many times. Consequently, there is often a greater chance of committing a Type II error than a Type I error. These different risks of error carry different ethical implications.

The normal practice in science-informed policy and management is to weigh the consequences of action versus making a Type I error. Under this condition, policy favors the interest of land development if the decision is to develop in light of scientific evidence that it will not damage the natural system. In turn, the burden is placed on the public, or regulatory agencies and the environmental movement that represent the public interest to argue for consideration of the potential harmful effects of development. It is the public and ultimately the ecosystem itself that assumes the risk of land development if the decision was faulty and there was a failure to safeguard nature.

If we change the ethical basis of decision-making by forcing policy and management to weigh the consequences of action versus making a Type II error, then we place the burden on the developer to show that land development will not jeopardize the public interest and ecosystem functioning. It is the developer that must assume the risk in this case. But, as I mentioned above, demonstrating that development is unlikely to inflict damage requires a considerable amount of experimentation—a costly undertaking. This then argues for taking a decidedly precautionary approach to tinkering with natural economies, which brings me back to the question that I posed in the introduction: Is dismissing a potential environmental problem for lack of understanding about clear cause a wise decision? I hope that the reader is coming to realize that the answer to this question should be a resounding "No". Dismissing for lack of cause means that decisions are being made based on risks of committing Type I errors. But, in suggesting that the answer should be "No" I do not imply that we should halt all human enterprises that have impacts on the environment. Intelligent tinkering is possible (chapter 7). But tinkering for sustainable function means that humans must be considered as integral parts of natural ecosystems (rather than outside their sphere) and that human impacts have manifold feedbacks that may eventually cost human welfare.

I have presented examples of the interdependent global links among species and ecosystems in chapter 7 and how acting within one ecosystem without regard to another can have devastating consequences. A much starker case in point is the reverberations caused by the collapse of the northwest Atlantic cod fishery and its feedbacks (current and future) on the human societies globally. The collapse of this species population due in part if not entirely to unsustainable exploitation by North American and

> Tinkering for sustainable function means that humans must be considered as integral parts of natural ecosystems (rather than outside their sphere) and that human impacts have manifold feedbacks that may eventually cost human welfare.

European Union fishing fleets has had devastating socioeconomic effects on coastal fishing communities in eastern Canada (Roughgarden and Smith 1996). Moreover, this fishery will take generations of human lives to recover to economic viability, if it recovers at all. This example illustrates how current exploitation and loss of species has limited the options for resource use by future generations.

The European Union fleets, in turn, have shifted their exploitation to the coasts of western Africa and now threaten to diminish fish stocks there. These fleets compete heavily with local Africans, especially in Ghana, who rely on fisheries for protein. Diminishing protein supplies from overfishing has increased demand for protein from other sources, primarily forest mammal species. This has lead to a lucrative but unsustainable bushmeat trade. Together, the exploitation of fisheries and mammal species has threatened many species populations in this region with imminent collapse. The consequences of collapse may be felt immediately in the form of widespread human poverty and food insecurity in the region (Brashares et al. 2004). With the diminishing stocks, European Union fleets may be forced once again to move elsewhere. It is easy and quite logical to reason that the effects of attendant, unsustainable fishing will eventually come full circle to impact Europeans' ability to meet their own protein demands.

> Taking a precautionary approach means that our definitions of prosperity must include maintaining a sustainable natural economy in addition to sustaining a market economy.

Precautionary approaches to resource exploitation require thinking through which interdependent species will ultimately be impacted and how this will alter the character and complexion of ecosystems when our children and grandchildren inherit the planet. It also recognizes that exploitation must be undertaken with sensitivity and respect for the livelihood and dignity of all human societies (Ludwig et al. 2001) and to minimize the risk of long-term negative consequences of the impacts. Taking a precautionary approach means that our definitions of prosperity must include maintaining a sustainable natural economy in addition to sustaining a market economy. It also means that ecologists, as scientists, must develop methods and approaches that better support precautionary decisions making in policy and management.

Policy and Management as a Scientific Enterprise

Basic science is essentially an adaptive process in which an initial hypothesis for the cause of a natural pattern or process is systematically tested. The hypothesis is either validated or rejected. In basic science, the hope is that the hypothesis is rejected because this typically leads to new and exciting discovery, which is after all the hallmark of success in basic science. The knowledge gained from discovery is used to refine existing ideas or develop new and better ones. These ideas are then subject to the next round of testing. This process of idea formulation, testing, discovery, and refinement is a time-tested cycle of knowledge creation.

Basic science in this purest of form, that is, accumulation of knowledge about nature for the simple sake of gaining knowledge, is neither intended nor geared toward solving specific applied environmental problems. Any scientific insights that are eventually used by policy are usually regarded as the dividend of the basic scientific process. Consequently, the interface between policy and science tends to be passive and linear. Basic science passes off to policy whatever knowledge is useful. Policy then uses this knowledge to formulate solutions to environmental problems.

The drawback of a linear, passive interface is that science continues with its cycle of discovery independent of the policy process. It is therefore entirely conceivable that policy could be based on outdated scientific knowledge by the time it has passed through the public review process and becomes implemented. The remedy is to intertwine science and policy. There are two creative ways to do this.

The first way is to change our perception about what policy and management is and what it can accomplish. Management is typically viewed as the implementation of policy and, as such, it alters some component of the natural environment to achieve a desired end. In essence, this is just a large-scale perturbation or treatment that is conducted in the absence of a control (Sinclair 1991). If management changed its approach slightly and left some areas unmanipulated (experimental controls), it can be treated as a scientific experiment that has the potential to offer knowledge about ecosystem function (Sinclair 1991). Let me illustrate this point with a hypothetical example of a lost opportunity because management did not include a control.

Northern forests in eastern North America are periodically threatened by the outbreak of insect pest species. Caterpillar stages of many moth species can grow to thousands and rapidly consume leaves of many eco-

nomically valuable deciduous and coniferous trees. One solution to this devastating environmental problem might be to aerially spray a pesticide (which
has been a practice in some parts of the region). Suppose a certain pesticide
has been chosen after clinical laboratory trials have demonstrated its effectiveness against the caterpillar pest.

There are, however, two problems with applying the pesticide based on
knowledge acquired from the clinical studies. First, a clinical trial cannot
guarantee against confounding random environmental effects when the pesticide is applied to a whole ecosystem. More importantly, many pesticides
are broad-spectrum agents, meaning that they kill more than the focal pest
species. Many pesticides have the potential to kill a whole host of butterflies
and moths that may offer important beneficial services in natural ecosystems
such as pollinating plants or being an important food base for many songbird species. Thus, a large-scale aerial application of the pesticide has the potential benefit of arresting the pest species outbreak. But the potential cost
of the application is damage to other important components of the northern forest ecosystem. Suppose that despite vocal public concerns, policy
makers felt that the risk was worth the market economic benefit of healthy
trees and thus called for the implementation of large-scale aerial spraying.
Suppose however, that management conducted a single, large-scale application of the pesticide. In so doing, it missed the opportunity to treat this as
an experiment with a treatment area and control area and follow-up monitoring of species and ecosystem responses to the pesticide.

Suppose that the application of the pesticide was confounded by a large-
scale, unavoidable random environmental effect after it was applied. For example, an unusually cool autumn and a prolonged rainy and cold spring
would be sufficient to cause a natural downturn in the pest species population via a large-scale reproductive failure. If such a random environmental effect occurred, it would swamp out the pesticide effect leading to a high
degree of uncertainty about the true effectiveness of the aerial spraying pro-
gram. The significance of this uncertainty is even more striking when we
consider it in light of the potential ancillary damage that the pesticide application may have caused to other ecosystem components. The absence of
a control in this case meant that management missed a valuable opportunity to collect important data on ecosystem-level responses to a large-scale
perturbation. Such information could have been used to devise more strategic and targeted pesticide applications in the future. More importantly, if
management applied the same approach to any later outbreak and justified

its action on the putative success realized here then it would be making an egregious Type II error.

The need to understand the risk of Type II errors brings us to the second way to blend science and policy—called active adaptive management (Walters and Holling 1990) or intelligent scientific tinkering. Active adaptive management treats a policy decision as a working hypothesis that contains logical predictions about outcome. If we implement the policy as a properly controlled management experiment, it is possible to evaluate the outcome systematically and discern fairly quickly whether or not the management is working. If not, we refine our thinking about management strategies, implement them, and carry out the next round of experimental evaluations of the new management regimes. Moreover, different management regimes (experimental treatments) can be carried out on smaller scales so we don't commit all of our eggs to one basket. Essentially, we "learn-by-doing" (Walters and Holling 1990) by capitalizing on the adaptive nature of the scientific process: Management is continually refined as we learn from successes (management regimes that come closest to achieving policy goals) and failures (management regimes that work less well). The important point is that failure should be celebrated, not punished. Failure, in this case of a planned experiment, is not a consequence of negligence (i.e., making decisions without considering the likelihood of Type II errors) but rather due to a comparison of different management regimes. It allows us to select the best performing management options and abandon the poorer performers.

In this vein, the role of the ecologist, as scientist, is not to advocate for one solution or another. Rather, ecologists contribute to policy by:

- Providing scientific insight on ecological interactions and the impacts of human activities on those interactions.
- Presenting the scientific insights in ways that reveal the trade-offs that different interest groups in the policy process must reconcile when making decisions.
- Illuminating the consequences of choosing one or the other trade-off option.

This is not to suggest that an ecologist, as an active citizen, cannot advocate for a preference. But, ethically the ecologist must make it clear whether he or she is acting as a scientist who is providing a balanced view of the science underlying the environmental issues, as opposed to advocating a certain policy based on personal preference. This is a difficult challenge to reconcile because, as was so poignantly stated by Aldo Leopold in 1953: " . . . the penalties of an ecological education is that one lives alone in a world of wounds." As ecologists, we see the hallmarks of unsustainable human activities and thus often wonder if, like the deer in the essay "Thinking Like a Mountain" (Chapter 7), we will be the next species to die because of our own "too much". I personally believe that the future is much brighter than that. But, getting there requires a realignment of ethical thinking in which market and natural economies are viewed as intertwined and interdependent. The good of each species is that it provides options for conserving the functioning of the whole ecosystem, even if our understanding of ecosystem function is incomplete.

Questions for Discussion

Chapter 1

1. What are three major issues confronting society in sustaining ecosystem functioning in the face of economic development?
2. Why are environmental problems increasingly challenging to resolve scientifically in modern society?

Chapter 2

1. What are the scientific aims of ecology?
2. What is an ecosystem?
3. Can you identify four different trophic levels in an ecosystem and explain their ecological role?
4. What do we mean by the term "biodiversity"?
5. What determines ecological complexity?
6. How do direct and indirect effects operate in ecosystems?
7. Species diversity in an ecosystem can be determined by predators and resources. How would you experimentally test between these two plausible causes of biodiversity? Be sure to state your hypotheses.
8. Can the scientific method of retroduction lead to reliable knowledge for policy?
9. What do we mean by the "game of life" in ecology?

Chapter 3

1. How does the greenhouse effect lead to warming of the planet? How do rising carbon dioxide levels influence the greenhouse effect?

2. What biophysical factors determine the location of specific ecosystems globally?
3. What happens when a species finds itself in environmental conditions that it cannot tolerate? In your answer, consider short-term (seasonal) and long-term (centuries) timescales.
4. What do we mean by climate envelope?
5. Can you identify two documented responses of species to global climate change?

Chapter 4

1. What is ecological carrying capacity and how is it measured? How does predation factor into carrying capacity estimates?
2. Why do populations of species not grow indefinitely in abundance?
3. Is there a "balance of nature"?
4. What is the implication of stable limit cycles and stochastic fluctuations for the management of species populations?

Chapter 5

1. What is the difference between age-structured and stage-structured population dynamics?
2. Can you explain the biological basis for the three qualitative shapes of survivorship curves?
3. Why do species such as turtles lay so many eggs over their lifetimes?
4. Can you identify three different kinds of life cycles?

Chapter 6

1. What is a species-area curve?
2. How would you experimentally test the effects of habitat fragmentation on two bird species: one that is a good competitor and one that is a good disperser? Be sure to state your hypotheses.
3. What is the difference between alpha diversity and beta diversity?
4. Are all rare species threatened with extinction? Why or why not?
5. What is extinction debt?
6. How can habitat fragmentation disrupt direct and indirect dependencies among species in ecosystems?

Chapter 7

1. How does the loss of top predators affect ecosystems?
2. What do we mean by a legacy effect? What is the timescale of ecological legacies?
3. How would one test the effects of restoring top predators to an ecosystem after a prolonged absence? State your hypothesis and deductions about the predator effects.
4. Can you identify two ways that ecosystems are linked across landscapes?

Chapter 8

1. How do species and ecosystems provide human kind important life-support services?
2. What factors contribute toward ecosystem stability?
3. How does biodiversity ensure against loss of ecosystem function?
4. What are the ecosystem conservation implications of rivet species versus redundant species?
5. Can you identify three nonmarket ecosystem services of biodiversity and explain why they are important to human economic well-being?

Chapter 9

1. Can you explain the rationale for biodiversity hotspot conservation?
2. Why might parks and protected areas only provide limited value for long-term biodiversity protection?
3. Can you design a conservation strategy that simultaneously allows economic development and protection of ecosystem services?
4. How do we implement parks and protected areas with other strategies to maintain dynamic landscapes?
5. What is a metacommunity process? What are the implications of this process for conservation?
6. How will global climate change affect the ability of parks and protected areas to meet their mandate of protecting biodiversity?

Chapter 10

1. Is dismissing a potential environmental problem for lack of understanding about clear cause a wise decision?

2. Can you articulate your own conservation ethic that ensures sustainable ecosystem function and economic opportunity for human kind?
3. What is the first precaution in intelligent tinkering with ecosystems?
4. How does uncertainty lead to two different kinds of errors in policy decisions?

References

Armsworth, P. R. and J. E. Roughgarden. 2003. The economic value of ecological stability. *Proceedings of the National Academy of Science USA* 100:7147–7151.

Barbour, A. G. and D. Fish. 1993. The biological and social phenomenon of Lyme disease. *Science* 260:1610–1616.

Bartholomew, G. 1977. Body temperature and energy metabolism. In M. S. Gordon, ed., *Animal Physiology: Principles and Adaptation*, 3rd ed. New York: MacMillan Publishing, 364–449.

Beissinger, S. R. and M. I. Westphal. 1998. On the use of demographic models of population viability in endangered species management. *Journal of Wildlife Management* 62:821–841.

Berger, J., P. B. Stacey, L. Bellis, and M. P. Johnson. 2001. A mammalian predator-prey imbalance: Grizzly bears and wolf extinction affect avian neotropical migrants. *Ecological Applications* 11:947–960.

Brashares, J. S., P. Arcese, M. K. Sam, P. B. Coppolillo, A. R. E. Sinclair, and A. Balmford. 2004. Bushmeat hunting, wildlife declines and fish supply in West Africa. *Science* 306:1180–1183.

Burns, C. E., K. M. Johnston, and O. J. Schmitz. 2003. Global climate change and mammalian species diversity in U.S. national parks. *Proceedings of the National Academy of Sciences USA* 100:11474–11477.

Cardinale, B. J., C. T. Harvey, K. Gross, and A. R. Ives 2003. Biodiversity and biocontrol: Emergent impacts of a multi-enemy assemblage on pest suppression and crop yield in an agroecosystem. *Ecology Letters* 6:857–865.

Carroll, C., R. E. Noss, P. C. Paquet, and N. H. Schumaker. 2004. Extinction debt of protected areas in developing landscapes. *Conservation Biology* 18:1110–1120.

Carson, R. 1962. *Silent Spring*. New York: Houghton Mifflin Co.

Caswell, H. 2001. *Matrix Population Models: Construction, Analysis, and Interpretation*. Sunderland, MA: Sinauer Associates.

Caughley, G. 1976a. The elephant problem—An alternative hypothesis. *East African Wildlife Journal* 14:265–283.

Caughley, G. 1976b. Wildlife management and the dynamics of ungulate populations. In T. H. Croaker, ed., *Applied Biology*. New York: Academic Press, 183–246.

Costanza, R., R. d'Arge, R. de Groot, S. Farber, M. Grasso, B. Hannon, K. Limburg, et al. 1997. The value of the world's ecosystem services and natural capital. *Nature* 387:253–260.

Cote, S. D., T. P. Rooney, J. P. Tremblay, C. Dussault, and D. M. Waller. 2004. Ecological impacts of deer overabundance. *Annual Review of Ecology, Evolution and Systematics* 35:113–147.

Courtois, R., J. P. Ouellet, C. Dussault, A. Gingras. 2004. Forest management guidelines for forest-dwelling caribou in Quebec. *Forestry Chronicle* 80:598–607.

Crooks, K. R. and M. E. Soulé. 1999. Mesopredator release and avifaunal extinctions in a fragmented system. *Nature* 400:563–566.

Crouse, D. T., L. B. Cowder, and H. Caswell. 1987. A stage-based population model for loggerhead sea turtles and implications for conservation. *Ecology* 68:1412–1423.

Crowder, L. B., D. T. Crouse, S. S. Heppel, and T. H. Martin. 1994. Predicting the impact of turtle excluder devices on loggerhead sea turtle populations. *Ecological Applications* 4:437–445.

Curran, L. M, S. N. Trigg, A. K. McDonald, D. Astiani, Y. M. Hardiono, P. Siregar, I. Caniago, et al. 2004. Lowland forest loss in protected areas of Indonesian Borneo. *Science* 303:1000–1003.

Daily, G. C. 1997. *Nature's Services: Societal Dependence on Natural Ecosystems.* Washington, D.C.: Island Press.

Davidson, J. 1938. On the growth of the sheep population in Tasmania. *Transactions of the Royal Society of South Australia* 62:342–346.

Easterling, D. R., B. Horton, P. D. Jones, T. C. Peterson, T. R. Karl, D. E. Parker, M. J. Salinger, et al. 1997. Maximum and minimum temperature trends for the globe. *Science* 277:364–367.

Edelstein-Keshett, L. 1988. *Mathematical Models in Biology.* New York: Random House.

Ehrlich, P. R. and A. H. Ehrlich. 1981. *Extinction: The Causes and Consequences of the Disappearance of Species.* New York: Random House.

Endler, J. 1986. *Natural Selection in the Wild.* Princeton, N.J.: Princeton University Press.

Ferraz, G., G. J. Russell, P. C. Stouffer, R. O. Bierregaard, S. L. Pimm, and T. E. Lovejoy. 2003. Rates of species loss from Amazonian forest fragments. *Proceedings of the National Academy of Science USA* 100:14069–14073.

Forbes, G. J. and J. B. Theberge. 1996. Cross-boundary management of Algonquin Park wolves. *Conservation Biology* 10:1091–1097.

Frank, D. A. and S. J. McNaughton. 1991. Stability increases with diversity in plant-communities: Empirical evidence from the 1988 Yellowstone drought. *Oikos* 62: 360–362.

Gates, D. M. 1980. *Biophysical Ecology.* New York: Springer-Verlag.

Gibbs, H. L. and P. R. Grant 1987. Oscillating selection on Darwin's Finches. *Nature* 327:511–513.

Hector A., B. Schmid, C. Beierkuhnlein, M. C. Caldeira, M. Diemer, P. G. Dimitrakopoulos, J. A. Finn, et al. 1999. Plant diversity and productivity experiments in European grasslands *Science* 286:1123–1127.

Helferich, G. 2004. *Humboldt's Cosmos: Alexander von Humboldt and the Latin American Journey that Changed the Way We See the World.* New York: Gotham Books.

Holdridge, L. R. 1947. Determination of world plant formations from simple climatic data. *Science* 105:367–368.

Holt, R. D. 2000. Trophic cascades in terrestrial ecosystems: Reflections on Polis et al. *Trends in Ecology and Evolution* 15:444–445.

Intergovernmental Panel on Climate Change (IPCC). 2001. *Climate Change 2001: Impacts, Adaptation, and Vulnerability.* Contribution of working group II to the third assessment report of the IPCC. New York: Cambridge University Press.

Jefferies, R. L., R. F., Rockwell, and K. E. Abraham. 2004. Agricultural food subsidies, migratory connectivity and large-scale disturbance in arctic ecosystems: A case study. *Integrative and Comparative Biology* 44:130–139.

Johnston, K. M. and O. J. Schmitz. 1997. Wildlife and climate change: Assessing the sensitivity of selected species to simulated doubling of atmospheric CO_2. *Global Change Biology* 3:531–544.

Kareiva, P. and U. Wennergren. 1995. Connecting landscape patterns to ecosystem and population processes. *Nature* 373:299–302.

Kennedy T. A., S. Naeem, K. M. Howe, J. M. H. Knops, D. Tilman, and P. Reich. 2002. Biodiversity as a barrier to ecological invasion. *Nature* 417:636–638.

Krebs, C. J. 1989. *Ecological Methodology.* New York: Harper and Row.

Kremen, C., N. M. Williams, and R. W. Thorp. 2002. Crop pollination from native bees at risk from agricultural intensification. *Proceedings of the National Academy of Science USA* 99:16812–16816.

Lawton, J. H. 1999. Are there general laws in ecology? *Oikos* 84:177–192.

Leibold, M. A., M. Holyoak, N. Mouquet, P. Amarasekare, J. M. Chase, M. F. Hoopes, R. D. Holt, et al. 2004. The metacommunity concept: A framework for multi-scale community ecology. *Ecology Letters* 7:601–613.

Leopold, A. 1953. *Round River.* Oxford, U.K.: Oxford University Press.

Levin, S. A. 1999. *Fragile Dominion: Complexity and the Commons.* Cambridge, MA: Perseus Publishing.

Levins, R. 1969. Some demographic and genetic consequences of environmental heterogeneity for biological control. *Bulletin of the Entomological Society of America* 15:237–240.

Likens, G. E. and F. H. Borman. 1974. Acid rain—Serious regional environmental problem. *Science* 184:1171–1179.

Lubchenco, J., A. M. Olson, L. B. Brubaker, S. R. Carpenter, M. M. Holland, S. P. Hubbell, S. A. Levin, et al. 1991. The sustainable biosphere initiative: An ecological research agenda. *Ecology* 72:371–412.

Ludwig, D., M. Mangel, and B. Haddad. 2001. Ecology, conservation and public policy. *Annual Review of Ecology and Systematics* 32:481–517.

Mack, R. N., D. Simberloff, W. M. Lonsdale, H. Evans, M. Clout, and F. A. Bazzaz. 2000. Biotic invasions: Causes, epidemiology, global consequences and control. *Ecological Applications* 10:689–710.

Mayr, E. 1982. *The Growth of Biological Thought: Diversity, Evolution and Inheritance.* Cambridge, MA: Harvard University Press.

McCann, K. S. 2000. The diversity-stability debate. *Nature* 405:228–233.

McKane R. B., L. C. Johnson, G. R. Shaver, K. J. Nadelhoffer, E. B. Rastetter, B. Fry, A. E. Giblin, et al. 2002. Resource-based niches provide a basis for plant species diversity and dominance in arctic tundra. *Nature* 415:68–71.

McShea, W. J., H. B. Underwoood, and J. H. Rappole. 1997. *The Science of Overabundance: Deer Ecology and Population Management.* Washington D.C.: Smithsonian Institution Press.

Myers, N. 1996. Environmental services of biodiversity. *Proceedings of the National Academy of Science USA* 93:2764–2769.

Myers, N., R. A. Mittermeier, C. G. Mittermeier, G. A. B. da Fonseca, and J. Kent. 2000. Biodiversity hotspots for conservation priorities. *Nature* 403:853–858.

Nee, S. and R. M. May. 1992. Dynamics of metapopulations—Habitat destruction and competitive coexistence. *Journal of Animal Ecology* 61:37–40.

Newmark, W. D. 1987. A land-bridge island perspective on mammalian extinctions in western North American parks. *Nature* 325:430–432.

Ostfeld, R. S., C. D. Canham, K. Oggenfuss, R. J. Winchcombe, and F. Keesing. 2006 Climate, deer, rodents, and acorns as determinants of variation in Lyme-disease risk. *Public Library of Science (PLoS) Biology* 4(6):e145.

Pacala, S. and G. Hurtt. 1993. Terrestrial vegetation and climate change: Integrating models and experiments. In P. Kareiva, J. Kingsolver, and R. B. Huey, eds., *Biotic Interactions and Global Change.* Sunderland, MA: Sinauer Associates, 57–74.

Paine, R. T. 1966. Food web complexity and species diversity. *American Naturalist* 100:65–73.

Parmesan, C. and G. Yohe 2003. A globally coherent fingerprint of climate change impacts across natural systems. *Nature* 421:37–42.

Parmesan, C., N. Ryrholm, C. Stefanescu, J. K. Hill, C. D. Thomas, H. Descimon, B. Huntley, et al. 1999. Pole-ward shifts in geographical ranges of butterfly species associated with regional warming. *Nature* 399(6736):579–583.

Peters, R. H. 1991. *Critique for Ecology.* New York: Cambridge University Press.

Peters, R. H. 1986. The role of prediction in limnology. *Limnology and Oceanography* 31:1143–1159.

Pianka, E. R. 1988. *Evolutionary Ecology*, 3rd ed. New York: Harper and Row.

Pimm, S. L. 1991. *The Balance of Nature?* Chicago: University of Chicago Press.

Porter, W. P. and D. M. Gates. 1969. Thermodynamic equilibria of animals with the environment. *Ecological Monographs* 39:245–270.

Post, E., R. O. Peterson, N. C. Stenseth, and B. E. McLean. 1999. Ecosystem consequences of wolf behavioral response to climate. *Nature* 401:905–907.

Rabinowitz, D., S. Cairns, and T. Dillon. 1986. Seven forms of rarity and their frequency in the flora of the British Isles. In M. E. Soulé, ed., *Conservation Biology: The Science of Scarcity and Diversity.* Sunderland, MA: Sinauer Associates, 182–204.

Ricciardi, A. and H. J. MacIsaac. 2000. Recent mass invasions of the Great Lakes by Ponto-Caspian species. *Trends in Ecology and Evolution* 15:62–65.

Ricklefs, R. E. and D. Schluter. 1993. *Species Diversity in Ecological Communities: Historical and Geographical Perspectives.* Chicago: University of Chicago Press.

Ripple, W. J. and R. L. Beschta. 2003. Wolf reintroduction, predation risk, and cottonwood recovery in Yellowstone National Park. *Forest Ecology and Management* 184:299–313.

Romesburg, H. C. 1981. Wildlife science: Gaining reliable knowledge. *Journal of Wildlife Management* 45:293–313.

Rosenzweig, M. L. 1995. *Species Diversity in Space and Time*. Cambridge, U.K.: Cambridge University Press.

Roughgarden, J. and F. Smith. 1996. Why fisheries collapse and what to do about it. *Proceedings of the National Academy of Science USA* 93:5078–5083.

Saberwal, V. K., J. P. Gibbs, R. Chellam, and A. J. T. Johnsingh. 1994. Lion-human conflict in the Gir Forest, India. *Conservation Biology* 8:501–507.

Schemnitz, S. D. 1980. *Wildlife Management Techniques Manual*, 4th ed. Washington, D.C.: The Wildlife Society Press.

Schindler, D. W. 1974. Eutrophication and recovery in experimental lakes: Implications for lake management. *Science* 184:897–899.

Schmidt, B., J. Joshi, and F. Schlapfer. 2001. Functional evidence for biodiversity-ecosystem function relationships. In A. P. Kinzig, S. W. Pacala, and D. Tilman, eds., *The Functional Consequences of Biodiversity*. Princeton, NJ: Princeton University Press, 120–168.

Schmitz, O. J. 2005. Scaling from plot experiments to landscapes: Studying grasshoppers to inform forest ecosystem management. *Oecologia* 145:225–234.

Schmitz, O. J. 2003. Top predator control of diversity and productivity in an old-field ecosystem. *Ecology Letters* 6:156–163.

Schmitz, O. J., E. Post, C. E. Burns, and K. M. Johnston. 2003. Ecosystem responses to global climate change: Moving beyond color-mapping. *BioScience* 53:1199–1205.

Schwartzman, S., A. Moreira, and D. Nepstad. 2000. Rethinking tropical forest conservation: Perils in parks. *Conservation Biology* 14:1351–1357.

Simberloff, D. S. and L. G. Abele. 1976. Island biogeography theory and conservation practice. *Science* 191:285–286.

Sinclair, A. R. E. 1997. Carrying capacity and the overabundance of deer. In W. H. McShea, H. B. Underwoood, and J. H. Rappole, eds., *The Science of Overabundance: Deer Ecology and Population Management*. Washington, D.C.: Smithsonian Institution Press, 380–394.

Sinclair, A. R. E. 1991 Science and the practice of wildlife management. *Journal of Wildlife Management* 55:767–773.

Sinclair, A. R. E., D. S. Hik, O. J. Schmitz, G. G. F. Scudder, D. H. Turpin, and N. C. Larter. 1995. Biodiversity and the need for habitat renewal. *Ecological Applications* 5:579–587.

Soulé, M. E., J. A. Estes, B. Miller, and D. L. Honnold. 2005. Strongly interacting species: Conservation policy, management, and ethics. *BioScience* 55:168–176.

Spencer, C. N., B. R. McClelland, and J. A. Stanford. 1991. Shrimp stocking, salmon collapse and eagle displacement. *BioScience* 41:14–21.

Speth, J. G. 2004. *Red Sky at Morning: America and the Crisis of the Global Environment*. New Haven, CT: Yale University Press.

Terborgh J. 2000. The fate of tropical forests: A matter of stewardship. *Conservation Biology* 14:1358–1361.

Thomas, C. D., A. Cameron, R. E. Green, M. Bakkenes, L. J. Beaumont, Y. C. Collingham, B. F. N. Erasmus, et al. 2004. Extinction risk from climate change. *Nature* 427:145–148.

Tilman, D. 1996. Biodiversity: Population versus ecosystem stability. *Ecology* 77:360–363.

Tilman, D., R. M. May, C. T. Lehman, and M. Nowak 1994. Habitat destruction and the extinction debt. *Nature* 371:65–66.

VEMAP Members. 1995. Vegetation/ecosystem modeling and analysis project (VEMAP): Comparing biogeography and biogeochemistry models in a continental-scale study of terrestrial ecosystem responses to climate change and CO_2 doubling. *Global Biogeochemical Cycles* 9:407–437.

Walker, B. 1992. Biodiversity and ecological redundancy. *Conservation Biology* 6:18–23.

Walters, C. J. and C. S. Holling. 1990. Large-scale management experiments and learning by doing. *Ecology* 71:2060–2068.

Worster, D. 1994. *Nature's Economy: A History of Ecological Ideas*, 2nd ed. Cambridge, U.K.: Cambridge University Press.

Yachi, S. and M. Loreau. 1999. Biodiversity and ecosystem productivity in a fluctuating environment: The insurance hypothesis. *Proceedings of the National Academy of Science USA* 96:1463–1468.

Zhou, Z. H. and W. S. Pan. 1997. Analysis of the viability of a giant panda population. *Journal of Applied Ecology* 34:363–374.

Further Reading

Chapter 1

Levin, S. A. 1999. *Fragile Dominion: Complexity and the Commons*. Cambridge, MA: Perseus Publishing.

Sala, O. E., F. S. Chapin, J. J. Armesto, E. Berlow, J. Bloomfield, R. Dirzo, E. Huber-Sanwald, et al. 2000. Global biodiversity scenarios for the year 2100. *Science* 287:1770–1774.

Chapter 2

Carpenter, S. R., S. W. Chisolm, C. J. Krebs, D. W. Schindler, and R. F. Wright. 1995. Ecosystem experiments. *Science* 269:324–327.

Golley, F. B. 1998. *A Primer for Environmental Literacy*. New Haven, CT: Yale University Press.

Krebs, C. J. 1987. *The Message of Ecology*. New York: HarperCollins Publishers.

Mayr, E. 2001. *What Evolution Is*. New York: Basic Books.

Weiner, J. 1994. *The Beak of the Finch: A Story of Evolution in Our Time*. New York: Knopf Publishing.

Chapter 3

Bazzaz, F. A. 1996. *Plants in Changing Environments: Linking Physiological, Population, and Community Ecology*. Cambridge, UK: Cambridge University Press.

Lovejoy, T. E. and L. Hannah. 2005. *Climate Change and Biodiversity*. New Haven, CT: Yale University Press.

Schmidt-Nielsen, K. 1998. *The Camel's Nose: Memoirs of a Curious Scientist*. Washington, D.C.: Island Press.

Intergovernmental Panel on Climate Change (IPCC). 2001. *Climate Change 2001: Impacts, Adaptation and Vulnerability*. Contribution of working group II to the third assessment report of the IPCC. New York: Cambridge University Press.

Walther, G. R., E. Post, P. Convey, A. Menzel, C. Parmesan, T. J. C. Beebee, J. M. Fromentin, et al. 2002. Ecological responses to recent climate change. *Nature* 416:389–395.

Chapter 4

Clutton-Brock, T. H., F. E. Guinness, and S. D. Albon. 1982. *Red Deer: Behavior and Ecology of Two Sexes*. Chicago: University of Chicago Press.

Cohen, J. E. 1995. *How Many People Can the Earth Support?* New York: W. W. Norton and Company.

Gotelli, N. J. 2001. *A Primer for Ecology*. Sunderland, MA: Sinauer Associates.

Sinclair, A. R. E. 1977. The African Buffalo: A Study of Resource Limitation of Populations. Chicago: University of Chicago Press.

Turchin, P. 2003. *Complex Population Dynamics: A Theoretical/Empirical Synthesis*. Princeton, NJ: Princeton University Press.

Chapter 5

Beissinger, S. R. and D. R. McCullogh. 2002. *Population Viability Analysis*. Chicago: University of Chicago Press.

Doak, D., P. Kareiva, and B. Kleptetka. 1994. Modeling population viability for the desert tortoise in the western Mojave desert. *Ecological Applications* 4:446–460.

Morris, W. F. and D. F. Doak. 2002. *Quantitative Conservation Biology: Theory and Practice of Population Viability Analysis*. Sunderland, MA: Sinauer Associates.

Chapter 6

Bierregaard, R., C. Gascon, T. E. Lovejoy, and R. Mesquita. 2001. *Lessons from Amazonia: The Ecology and Conservation of a Fragmented Forest*. New Haven, CT: Yale University Press.

Hanski, I. 1999. *Metapopulation Ecology*. Oxford, UK: Oxford University Press.

Schmiegelow, F. K. A., C. S. Machtans, and S. J. Hannon. 1997. Are boreal birds resilient to forest fragmentation? An experimental study of short-term community responses. *Ecology* 78:1914–1932.

Terborgh, J., L. Lopez, P. Nunez, M. Rao, G. Shahabuddin, G. Orihuela, M. Riveros, et al. 2001. Ecological meltdown in predator-free forest fragments. *Science* 294:1923–1926.

Chapter 7

Leopold, A. 1966. The land pyramid. In *A Sand County Almanac with Other Essays on Conservation from Round River*. Oxford, UK: Oxford University Press, 230–236.

Pauly, D., V. Christensen, J. Dalsgaard, and R. Froese. 1998. Fishing down marine food webs. *Science* 279:860–863.

Polis, G. A., M. E. Power, and G. R. Huxel. 2004. *Food Webs at the Landscape Level*. Chicago: University of Chicago Press.

Sinclair, A. R. E. and A. E. Byrom. 2006. Understanding ecosystem dynamics for conservation of biota. *Journal of Animal Ecology* 75:64–79.

Soulé, M. E., J. A. Estes, J. Berger, and C. M. Del Rio. 2003. Ecological effectiveness: Conservation goals for interactive species. *Conservation Biology* 17:1238–1250.

Chapter 8

Daily, G. C. and K. Ellison. 2002. *The New Economy of Nature: The Quest to Make Conservation Profitable.* Washington, D.C.: Island Press.

Kinzig, A. P., D. Tilman, and S. Pacala. 2001. *The Functional Consequences of Biodiversity: Empirical Progress and Theoretical Extensions.* Princeton, NJ: Princeton University Press.

Naeem, S., F. S. Chapin III, R. Costanza, P. R. Ehrlich, F. B. Golley, D. U. Hooper, J. H. Lawton, et al. 1999. Biodiversity and ecosystem functioning: Maintaining natural life support processes. Issues in Ecology 4. Washington D.C.: Ecological Society of America. At www.esa.org/science/Issues/TextIssues/issue4.php.

U.S. National Academy Committee on Noneconomic and Economic Value of Biodiversity. 1999. *Perspectives on Biodiversity: Valuing Its Role in an Everchanging World.* Washington, D.C.: National Academy Press.

Chapter 9

Fischer, J., D. B. Lindenmayer, and A. D. Manning. 2006. Biodiversity, ecosystem function, and resilience: ten guiding principles for commodity production landscapes. *Frontiers in Ecology and the Environment* 4:80–86.

James, A., K. J. Gaston, and A. Balmford. 2001. Can we afford to conserve biodiversity? *BioScience* 51:43–52.

Lubchenco, J., S. R. Palumbi, S. D. Gaines, and S. Andelman. 2003. Plugging a hole in the ocean: The emerging science of marine reserves. *Ecological Applications* 13 (supplement):S3–S7.

Sinclair, A. R. E., D. Ludwig, and C. W. Clark. 2000. Conservation in the real world. *Science* 289:1875–1875.

Stolton, S. and N. Dudley. 1999. *Partnerships for Protection: New Strategies for Planning and Management for Protected Areas.* London: Earthscan Publications.

Chapter 10

Callicott, J. B. 1999. *Beyond the Land Ethic: More Essays in Environmental Philosophy.* Albany: State University of New York Press.

Ehrlich, P. R. and A. H. Ehrlich. 2004. *One with Nineveh: Politics, Consumption, and the Human Future.* Washington, D.C.: Island Press.

Kareiva, P. 2005. Is the key to conservation changing ethical values or policing unethical behavior? *Current Biology* 15: R40–R42.

Leopold A. 1966. The land ethic. In *A Sand County Almanac with Other Essays on Conservation from Round River.* Oxford, UK: Oxford University Press, 217–240.

Glossary

Adaptation: The evolution, via natural selection, of organisms' traits. These include everything from physiological coping mechanisms to predator hunting tactics.

Alpha diversity (α diversity): A measure of the variety of species within a local area or habitat.

Balance of nature: A vernacular term used to describe the condition in which natural checks such as predation or starvation compensate for births, thereby maintaining species populations in a constant state.

Beta diversity (β diversity): A measure of the variety of species across a landscape gradient comprised of several habitats.

Biodiversity: The biological variety at all levels of organization including genetic variety within a species population, and species variety within an ecological community.

Biodiversity hotspot: A geographic region in which there is an extraordinarily large concentration of species that evolved in that region.

Carrying capacity: The capacity of a habitat or area to support a population at fixed numbers through the supply of resources necessary for survival and replacement reproduction.

Climate-space envelope: A mathematical construct that describes the combinations of solar radiation and environmental temperatures that a species can tolerate based on its physiological coping mechanisms.

Complex life cycle: Life cycle development in which offspring are altogether functionally different than adults. A classic example is the development of butterflies, which follows from egg to larva feeding on vegetation to resting pupa to adult feeding on nectar and pollen.

Deterministic population growth: Population growth that can be accurately predicted because the birth and death rates, and immigration and emigration rates remain constant over time.

Direct effect: The immediate impact of one organism on another's chance of survival or reproduction through a physical interaction such as predation or interference.

Diversity-stability: The relationship that exists when the likelihood of long-term sustainability of an ecological system increases with the diversity of species interconnections within that system.

Ecosystem: A descriptor of a biological system on the basis of the complex of species inhabiting a region as well as the chemical and physical attributes of that region.

Ecosystem stability: The capacity of an ecosystem to maintain or return to normal species abundances or function in the face of disturbances.

Ecosystem resilience: A form of ecosystem stability: it is measured as the rate at which an ecosystem returns to normal function or species abundances following a disturbance.

Ecosystem resistance: A form of ecosystem stability: it is measured as the capacity of an ecosystem to remain unchanged in the face of a disturbance.

Equilibrium: The ecological condition in which death or emigration rates of species exactly balance birth or immigration rates.

Exponential growth: A form of population growth in which there is a multiplicative or compounded increase in the number of individuals over time.

Extinction debt: The compounded future loss of species diversity propagated by contemporary disruption of dominant species interactions in an ecosystem through habitat destruction.

Extraction reserve: A biodiversity conservation tool in which local societies are allowed to continue with traditional resource harvesting within a protected area in order to promote sustainable economies and ecosystems.

Fitness: The net lifetime contribution of an individual to the future genetic pool of a population, quantified in terms of its capacity to survive and reproduce.

Food chain: A descriptor of an ecological system in terms of the feeding linkages or energy flow among major groups of species, e.g., in a two-link chain, carnivores consume herbivores, which in turn consume plants.

Food web: The interconnected network of feeding linkages among species in a system.

Functional complementarity: The condition in which species bolster each others' roles in an ecosystem to the extent that their net combined effect is greater than their individual effects. For example, two predator species together can be more effective than each species individually when each predator feeds on different life cycle stages of their shared prey.

Global climate change: The change in the earth's climate brought about by rising atmospheric levels of gases from air pollution.

Greenhouse effect: The warming of the earth's surface due to absorption of infrared radiation by a gaseous atmosphere that metaphorically acts like glass windows in a greenhouse.

Greenhouse gas: Atmospheric gases that cause a greenhouse effect by creating a boundary layer around the earth that traps heat energy. These gases include water vapor, carbon dioxide, ozone, methane, and nitrous oxides.

Habitat fragmentation: The parceling of contiguous habitat area into smaller, isolated units consequent to natural disasters or land development.

Hypothetico-deduction: A scientific method in which an hypothesis is proposed and systematically tested with empirical data. The aim of this method is to reveal cause-effect relationships in nature.

Indigenous reserve: A biodiversity conservation tool in which local societies are allowed to continue with their traditional way of life within a protected area in order to harmonize conservation and development.

Indirect effect: The impact of one organism on another's chance of survival or reproduction mediated through a direct interaction with a third-party species. For example, a predator can indirectly enhance plant survival by directly preying on herbivores that feed on the plant.

Induction: A scientific method in which one identifies an empirical trend between two or more variables.

Keystone predator: A top predator species that has a dramatic effect on the species diversity in an ecosystem by controlling the abundance of strongly competitive species.

Lag effect: The condition in population dynamics when a variable (e.g., population density) is correlated with the values of that variable several time steps in the past.

Mesopredators: Middle-size predators in a food web that are preyed upon by large top predators.

Metapopulation: The set of spatially separated populations of the same species that are linked to each other through dispersal.

Metacommunity: The set of spatially separated communities of multiple interacting species that are linked by dispersal of those species.

Natural selection: A process in which individual differences in survival and reproductive success lead to evolutionary change.

Nonadditive effect: A condition in which the effect of two or more species on an ecosystem is unequal to the sum of the individual species' effects.

Population oscillation: The condition in which the abundance of a population cycles up and down over time.

Population viability analysis: An analysis of a population's risk of extinction based the demographic attributes (e.g., birth and death rates) of the population.

Redundant species: Species that have identical functional roles in an ecosystem, e.g., two herbivores that eat the same plant material.

Retroduction: A scientific method that ascribes a reason or hypothesis for the trend between two or more variables identified from induction.

Rivet species: Species that have unique but complementary functional roles in an ecosystem. The term derives from the metaphor of rivets in an airplane wing in which each rivet holds together a small part of the wing but together the rivets complement each other by maintain the wing's integrity.

Species–area curve: The mathematical relationship between the size of an area or habitat and the number of species contained within that area.

Species diversity: The variety of species in a location.

Species evenness: The degree to which species are equally represented in a location.

Stable limit cycle: A dynamic in which the population abundance of a species oscillates in a regular, repeatable manner for an indefinite time period.

Stochastic population growth: Population growth that cannot be accurately predicted because the birth and death rates and immigration and emigration rate are influenced strongly by random environmental changes.

Structured population: A population that is described in terms of its component age classes, e.g., newborn, juvenile, subadult, adult.

Survivorship curve: A mathematical relationship between the likelihood of surviving and age.

Sustainability: A vernacular term to describe long-term stability of ecosystem function.

Tolerance curve: A mathematical relationship between an organism's fitness and an environmental variable such as temperature or solar radiation.

Trophic interaction: A feeding interaction in a food chain or food web.

Trophic level: A level in a food chain to which a particular feeding group belongs, e.g., herbivores, carnivores, decomposers, etc.

About the Author

Oswald J. Schmitz is the Oastler Professor of Population and Community Ecology in the Yale School of Forestry and Environmental Studies. His research examines how species interactions in food webs determine the nature and level of ecosystem functions.

Index

Italicized page numbers refer to boxes and figures.

Active adaptive management, 137
Adaptation, 18–19, 33–35, 111, 124, 153
Aerial spraying, 136
Age–structured population dynamics modeling, 67, 68–75, *69, 70, 74*
Agriculture, 83–84, 97–98, 105, 111–13, 115, 118
Alfalfa production, 113
Algonquin Park (Ontario), 118–19
Alpha diversity, 79–80, 121, 153
Amphibian species, 44, 64, 66, 71, 116
Apex predators. *See* Keystone predators
Aphids, 113
Aquatic food webs, 99
Arctic ecosystems, 12, 79, 98
Aspen (*Populus tremuloides*), 102–3, *104*
Assortative mating, 35
Atlantic cod fishery, 133–34
Attractor, 56–57

Balance of nature, 49, 52, 61–62, 153. *See also* Equilibrium
Bald eagle (*Haliaeetus leucocephalus*), 96
Balsam fir (*Abies balsamea*), 100–101
Baobab trees, 54
Bears, 61, 93
Bee species, 112–13
Beta diversity, 80, 122–23, 153
BIODEPTH, 111
Biodiversity, 2, 4, 153; and climate–space envelope, 35–36, *37–38*, 40, *41*; and conservation ethic, 130–34; coping strategies with climate, 29–35, *33*; and ecosystem services, 102–14, *104, 107*; and ecosystem types, 28–29, *30*; effects of global climate change on, 36, 38–44; and habitat fragmentation, 79–91, *81, 87*; physics underlying, 26–28, *29*; protection of, 115–25; and web of life, 94, 99
Biodiversity hotspots, 115–17, 122, 153
Biological differences, 17–18

Biomass production, 105, 107, 109, 110–11

Biosphere reserves. *See* Indigenous reserves

Bird species, 42–44, 71, 84–88, 87, 94, 116, 129–32, 136

Boreal forest regeneration, 102–5, *104*

Buffer zones, 121

Butterfly species, 43, 66, 136

Canadian Group of Seven, 118

Carbon dioxide (CO_2), 27, 38, 40, *41*, 43, 100, 110

Cardinale, B. J., 113

Carrying capacity, 49–54, *50*, *53*, 56–62, *60*, 153

Carson, Rachel, 1

Caughley, Graeme, 45, 52, 54

Cause-effect basis, 20–22, 39, 57, 77–78, 79, 86, 131

Cedar Creek Long–Term Ecological Research site (Minn.), 107, 114

Chaparral habitat, 85–86, *87*, 90, 129–30. *See also* Sage–scrub habitat

Clean Water Act, 1

Clear–cut harvesting, 102, *104*

Climate: and climate–space envelope, 35–36, *37–38*, 40, *41*; coping with, 29–35, *33*; and ecosystem types, 28–29, *30*; effects of global change, 36, 38–44; physics underlying, 26–27, *29*. *See also* Global climate change; Weather

Climate–space envelope, 35–36, *37–38*, 40, *41*, 153

Coefficient of variation (CV), 107, *107*

Columbian ground squirrel (*Spermophilus columbianus*), 39, 42

Competition, 8–10, *9*, *11*, 12–15; and biodiversity protection, 122, 125; and ecosystem services, 103, *104*, 114; and habitat fragmentation, 82–83, 85, 89–90; and population size, 48–51, *50*, 54–55; and web of life, 95, 100

Complex life cycle, 153

Computer simulations, 39, 72, 75–76

Conservation: and biodiversity protection, 115, 116–21, *117*, 122; and extraction reserves, *117*, 120–21; and habitat fragmentation, 79–80, 82–84, 89; and indigenous reserves, *117*, 120–21; parks/protected areas, 116–19; and web of life, 99

Conservation ethic, 5, 92–93, 126–38; and intelligent tinkering, 126, 127–30, 133, 137; and policy/management as scientific enterprise, 135–38; questions concerning, 141–42; and scientific uncertainty/precaution, 130–34

Consumer–resource interactions, 8–10, *9*, *11*, 103, *104*, 107

Contest competition. *See* Interference competition

Continental scales, 97–98

Control, experimental, 24–25

Coping strategies, 17, 19; with climate, 29–35, *33*

Correlational basis. *See* Proximate basis

Cottongrass species (*Eriophorum vaginatum*), 12–13

Cottonwoods (*Populus* sp.), 94–95

Coyote (*Canis latrans*), 61, 85–86, *87*, 96

Crooks, K. R., 86

Crop pollination, 105, 111–13

Damped oscillation, 51–52, *53*

Damsel bug (*Nabis* sp.), 113

Darwin, Charles, 18

Darwinian fitness. *See* Fitness

Deer: and biodiversity protection, 118; and ecosystem services, 103; and global climate change, 29, 31, 39, 42; and Lyme disease, 21; and population size, 45, 59–62, *60*; and web of life, 92–93

Deer mice (*Peromyscus* sp.), 21

Deforestation, 2

Density–dependent growth, 49, *53*, 55–56
Density–independent growth, 49
Deterministic population growth, 55–57, 153
Direct effects, 9–15, *9*, *11*, 154; and global climate change, 39–40, *41*; and web of life, 94, 100–101
Dispersing species, 89–90
Disturbances: and conservation ethic, 135–36; and ecosystem services, 106–8, *107*; and habitat fragmentation, 82–83, 89; and population size, 52; sensitivity of populations to, 72–75, *74*; and web of life, 100–101
Diversity indices, 80, 82–83
Diversity–Stability Hypothesis, 106–9, *107*, *108*, 154
Drought, 18–19, 107–8
Dynamic landscapes, 121–24

Eastern chipmunk (*Tamias striatus*), 39, 42
Ecological Society of America, 3–4
Ecological succession, 122–23
Ecology, science of, 6–25; and conservation ethic, 130–34; definition of, 7; gaining reliable knowledge, 20–25; life-as-a-game metaphor, 15–20, 46; questions concerning, 139; resolving complexity, 7–15
Economic development, 2, 117, *117*, 120–21, 131–33
Economic incentives, 121
Ecosystem functions, 2, 105–14; and crop pollination, 105, 111–13; and ecosystem stability, 106–10, *107*, *108*; and invasion resistance, 105, 114; and pest control, 105, 113–14; and production costs, 105, 110–11; protection of, 115–25
Ecosystem resilience, 105, *107*, 154
Ecosystem resistance, *107*, 108–9, 114, 154
Ecosystems, 28–29, *30*, 154; in space, 96–101; in time, 93–96

Ecosystem services, 2, 102–14, *104*, 121; and boreal forest regeneration, 102–5, *104*; and conservation ethic, 126–27; and ecosystem functions, 106–14, *107*, *108*; and material goods, 105, 114; questions concerning, 141
Ecosystem stability, 105, 106–10, *107*, *108*, 154
Elephant (*Loxodonta africana*), 52, 54
Elk (*Cervus elaphus*), 35–36, 39–40, *41*, *42*, 94–95, 128
El Niño, 18–19
Endemic species, 116
Environmental advocacy, 61, 98
Equilibrium, 49–52, *50*, 55–56, 123, 154. *See also* Carrying capacity
European honey bee (*Apis mellifera*), 112
European Union fleets, 134
Evenness indices, 80, 82
Evolutionary fitness. *See* Fitness
Evolutionary processes, 15, 17–20, 34–35, 44, 71–72
Experimental perturbation, 24–25
Exploitative competition, 10, *11*, 12–13
Exponential growth, 46–49, *47*, 154
Extinction debt, 89–91, 118–19, 154
Extinctions, 2; and biodiversity protection, 115, 118–19, 121–23; and conservation ethic, 128–29; and global climate change, 44; and habitat fragmentation, 84–85, 88–91
Extraction reserves, *117*, 120–21, 154
Extrinsic limiting factors, 54–56, 61

Feedback cycles, 19
Ferraz study, 84–85
Fisheries, 76–78, 95, 133–34
Fishing trawler nets, 76–78
Fish species, 95, 99
Fitness, 16–17, *18*, 154; and global climate change, 31–32, *33*, 34–36, 40; and population size, 46, 48–49, 51, 54; and web of life, 99

Food chains, 7–8, 10, *11*, 12, 154; and ecosystem services, 106, *108*, 111; and web of life, 96

Food webs, 8–10, 9, *11*, 15, 154; and conservation ethic, 129; and ecosystem services, *104*, 106–7, *108*, 109–10; and habitat fragmentation, 85–86, *87*; and population size, 54; and web of life, 93–94, 99

Forest industry, 102–3, *104. See also* Logging

Frank, D. A., 107–9

Functional complementarity, 110, 111, 113, 128, 154

Functional questions, 20, 34

Functional Redundancy Hypothesis, 110

Galapagos Islands, 18–19

Game of life. *See* Life-as-a-game metaphor

Game species, 92–93

General Circulation Models (GCMs), 38, 40

Genetic transmission, 18, 46

Geographic ranges, 82–83, 96–101, 124–25

Geometric growth. *See* Exponential growth

Giant Panda (*Ailuropoda melanonleuca*), 72–73, *74*

Global climate change, 4, 19, 36, 38–44, 154; and biodiversity protection, 124–25; questions concerning, 139–40; and web of life, 100–101

Globalization, 98–100

Goldenrod (*Solidago rugosa*), 14

Government policies, 84, 93–94, 98, 135–38

Graded protection, 121

Grasshopper herbivore (*Melanoplus femur-rubru*), 14

Grass species (*Poa pratensis*), 14

Gray fox (*Urocyon cinereoargenteus*), 85, *87*

Greenhouse effect, 27, 124, 154

Greenhouse gases, 27, 36, 100, 155

Grizzly bear (*Ursus arctos*), 93, 96

Group of Seven, 118

Gulls (*Larus* sp.), 96

Habitat fragmentation, 2, 19, 43, 79–91, 155; and biodiversity protection, 118; and conservation ethic, 129, 131–32; and diversity indices, 80, 82–83; and ecosystem services, 109; and population/community processes, 88–91; questions concerning, 140; and species–area relationship, 79–80, *81*, 83–88, *87*

Habitat renewal, 123

Habitat specificity, 82–83

Harvested species, 45, 102, *104*, 109

Hawksbill turtles, 63–64

Herring gulls, 16

Heterogeneity indices, 80–81

Humboldt, Alexander von, 83

Humpty Dumpty effects, 129–30

Hunting, 61, 103, 118

Hunting spider carnivore (*Pisaurina mira*), 14

Hypotheses, 23–24, 131–32, 135, 137

Hypothetico–deduction, 24–25, 39, 155

Indigenous reserves, 117, *117*, 119, 120–21, 155

Indirect effects, 9–15, 9, *11*, 155; and global climate change, 39–40, 42–43; and habitat fragmentation, 85–86, *87*; and web of life, 93–94, 100–101

Induction, 22–23, 86, 130, 155

Infrared radiation, 27

Insect species, 44, 66, 71, 96–97, 135–36

Instar stages, 66

Intelligent tinkering, 126, 127–30, 133, 137

Interference competition, 10, *11*, 13

Intraguild effect, 113

Intrinsic limiting factors, 54

Invariant causal link, 79
Invasive species, 98–100, 105, 114
Island ecosystems, 96–97

Keystone predators, 14–15, *104*, 119, 155
Knowledge creation, 135
Kokanee salmon (*Onchorynchus nerka*),
    95–96

Labrador tea (*Ledum palustre*), 13
Ladybird beetle (*Harmonia axyridis*), 113
Lag effects, 51–52, 57, 59, 89, 93, 155
Lake trout (*Salvelinus namaycush*), 96
Land use patterns, 19, 43, 112, 116, *117*, 119,
    121, 125
"Learn-by-doing," 137
Leopold, Aldo, 92–93, 126–27, 130, 138
Life-as-a-game metaphor, 15–20, 46,
    64–65, 89, 99
Life cycles, 66–68, *67*
Limit cycle hypothesis, 53–54, 53
Loggerhead sea turtles (*Caretta caretta*),
    64–68, 65, 67, *69*, 75–78, 77
Logging, 83–84, 102–3, *104*, 111, 118
Low–bush cranberry (*Vaccinium vitis-
    idaea*), 12–13
Lyme disease, 21

MacArthur, Robert, 106–7
Management: and biodiversity protection,
    123; and conservation ethic, 128,
    131–32, 135–38; and ecosystem services,
    102–3, *104*, 112–13; and habitat frag-
    mentation, 79–80; and population size,
    45, 52, 54, 57–62, 60; and population vi-
    ability analyses, 63–64, 71–73, *74*, 75–78,
    77; and web of life, 93–95, 98, 101
Market economies, 2, 5, 127–28, 134, 138
Material goods, 105, 114
McNaughton, S. J., 107–9
Mesopredators, 85–88, *87*, 129, 155
Metabolism, 31–32, 36

Metacommunity, 121–22, 155
Metapopulation, 88–89, 155
Migration, 88–89, 98, 118–19, 122, 125
Migratory bird species, 94, 98
Mink (*Mustela vison*), 96
Mobility, 89
Models, 39–40, 42; age–structured popula-
    tion dynamics, 68–75, *69*, *70*, *74*
Monitoring, 77–78, 117, 136
Moose (*Alces alces*), 93–94, 100–101, 103,
    104, 118

National parks, 116–19, *117*, 121, 123–25
Native species, 99–100, 112–14, 116
Natural balance. *See* Balance of nature;
    Equilibrium
Natural economies, 2, 6; and conservation
    ethic, 127–30, 133–34, 136, 138; and
    ecosystem services, 105, 109, 111, 114
Natural selection, 17–20, 34–35, 44, 46,
    64, 155
Nonadditive effect, 155
North Atlantic Oscillation (NAO), 100
Null hypothesis, 131–32

Oceanic island ecosystems, 96–97
Opossum (*Didelphis virginanus*), 85, *87*
Opossum shrimp (*Mysis relicta*), 95
Opuntia cactus, 96
Organic farming, 112–13
Otter (*Lutra canadensis*), 96
Overabundance, 45, 52, 57–62, *60*
Overfishing, 134
Ozone layer, 27

Paradox of diversity, 12
Parasitic wasp (*Aphidius ervi*), 113
Pea aphid (*Acyrthosiphon pisum*), 113
Pest control, 1, 105, 113–14, 135–36
Phenotype, 35–36, 46
Photosynthesis, 29, 110
Physics underlying biodiversity, 26–28, *29*

Phytoplankton, 99–100
Plankton species, 95, 99–100
Policy, 84, 93–94, 98, 131–32, 135–38
Polis, Gary, 96
Pollination, 105, 111–13, 136
Pollution, 19, 100
Population compression hypothesis, 52, 54
Population eruption hypothesis, 52, 54
Population oscillation, 51–52, 53, 55–57, 155
Population processes, 88–91
Population size, 45–62; carrying capacity, 49–54, 50l, 53, 57–62, 60; and competition/predation, 50, 50, 54–55, 60, 61–62; and habitat fragmentation, 82–83, 88–91; overabundance, 45, 52, 57–62, 60; population growth, 46–49, 47; and population viability analyses, 75; questions concerning, 140; and weather, 55–57
Population viability analyses, 63–78; Giant Panda (Ailuropoda melanonleuca), 72–73, 74; and life cycles, 66–68, 67; loggerhead sea turtles (Caretta caretta), 64–68, 65, 67, 69, 75–78, 77; modeling age–structured dynamics, 68–75, 69, 70, 74
Population viability analysis, 155; and life cycles, 66–68
Precautionary approaches, 133–34
Precipitation, 28–29, 30
Predation, 8–10, 9, 11, 12–16; and biodiversity protection, 118–19, 122, 125; and conservation ethic, 128–30; and ecosystem services, 104, 113–14; and habitat fragmentation, 82–83, 85, 87, 90; and population size, 50, 50, 54–55, 60, 61–62; in population viability analyses, 63–64, 71; and web of life, 92–97, 99, 100
Predictive reliability, 79–80
Primrose (Primula scotia), 83
Priority effects, 129–30

Productivity, 105, 110–11
Protecting biodiversity, 115–25; and dynamic landscapes, 121–24; and global climate change, 124–25; and graded protection, 121; and hotspot strategy, 115–17, 122; and indigenous or extraction reserves, 117, 117, 119, 120–21; and parks/protected areas, 116–19; questions concerning, 141
Protein, 134
Proximate basis, 20

Questions for discussion, 139–42

Rabbit species, 34, 36
Raccoon (Procyon lotor), 85, 87
Rare species, 80, 82–83
Recolonization, 88–90, 118
Redundant species, 109–10, 111, 112, 128, 155
Regeneration, 102–3, 104, 105, 122
Reintroductions, 94–95, 128–30
Reptile species, 64, 71
Reshuffling of faunas, 124–25
Resilience. See Ecosystem resilience
Resistance. See Ecosystem resistance
Retroduction, 23–24, 64, 156
Riparian vegetation, 93–95
Rivet Hypothesis, 110
Rivet species, 109–10, 111, 112, 156

Sage-scrub habitat, 85–86, 87, 90, 129–30
Salmon, 95–96
Scientific methodology, 22–25, 111, 128, 130–32, 135–38
Scramble competition. See Exploitative competition
Sedge species (Carex bigelowii), 12
Seed-eating finches (Geospiza species), 18–19
Sensitivity analyses of structured population models, 72–75, 74

Sex ratio, 69–70
Shannon–Weiner index, 80
Sheep, Tasmanian, 47, 49, 51
Shrub dwarf birch (*Betula nana*), 13
Silent Spring (Carson), 1
Simpson's index, 80
Solar radiation, 26–28, *29*. See also Climate
Soulé, M. E., 86
Specialist species, 110
Species–area curve, 79–80, *81*, 156
Species–area relationships, 79–80, *81*,
    83–88, *87*, 118
Species diversity. *See* Biodiversity
Species evenness, 80, 82, 156
Species richness, 80, 84, 86, 108, 111,
    114, 130
Species–specific thresholds, 88
Spiders, 96–97
Stable limit cycles, 52–54, *53*, 156
Stable systems, 106. *See also* Ecosystem
    stability
Stage-structured population dynamics,
    67–68, 75
Starfish (*Pisaster ochraceus*), 14
Stochastic population growth, 56–57, 156
Stress, physiological, 16
Striped skunk (*Mephitis mephitis*), 85, *87*
Structured population, 156
Subsistence economies, 120
Successional stages, 122–23
Survivorship curve, 70–73, *70*, 156
Sustainability, 4, 6, 105, 109, 111, 120–21,
    128, 134, 156
"The Sustainable Biosphere Initiative"
    (ESA), 3–5

Temperature, 26–28, *29*. See also Climate
Temporal dynamics of habitats, 122–24
Thermodynamics, laws of, 127
"Thinking Like a Mountain" (Leopold),
    92, 138

Threatened species, 45, 63–78, 140
Tolerance, 31–32, *33*, 34, 124
Tolerance curve, 156
Toxic chemicals, 100
Transport of species, 98–100
Trawl fisheries, 76–78
Trophic cascades, 12, 14, 93, 95, 97–98,
    100, 129
Trophic levels/interactions, 8–10, *9*, *11*, 12,
    14, 156; and biodiversity protection,
    125; and conservation ethic, 129; and
    ecosystem services, 103, *104*, 106–9,
    108; and web of life, 93–95, 97–100
Tropical forests, 2, 79, 83–85, 89–90, 120
Turtle excluder devices (TEDs), 76–78
Type I error, 132–33
Type II error, 132–33, 137

Ultraviolet rays, 27
Uncertainty, 130–32, 136
Unstable systems, 106, 109. *See also*
    Ecosystem stability

Value trade-offs, 59–62, *60*
VEMAP project, 40

Weather, 18–19, 55–57, 100–101, 107–8.
    See also Global climate change
Web of life, 92–101; ecosystems in space,
    96–101; ecosystems in time, 93–96;
    questions concerning, 141
Whitefish (*Coregonus clupeaformis*), 95
White pine (*Pinus strobes*), 118
White spruce (*Picea glauca*), 102–3, *104*
White-tailed deer (*Odocoileus virginianus*),
    21, 29, 31, 39, 42, 45, 59–62, 60,
    103, 118
Wicked problems, 20
Wildlife species: forecasting global climate
    change effects on, 39–43. *See also* Bio-
    diversity; *names of individual species*

Wolf (*Canis lupus*), 61, 92–95, 100, 118–19, 128

Woodland caribou (*Rangifer tarandus caribou*), 82

Worster, Donald, 1

Yellowstone National Park, 94, 107–8, 128

Zebra mussel (*Dreissena polymorpha*), 99–100

Zooplankton species, 95, 99